A field guide to the

PINES

of Mexico and Central America

This work was undertaken at the Oxford Forestry Institute, University of Oxford, on behalf of the Overseas Development Administration, under project number R5465 of the Forestry Research Programme.

A field guide to the PINES of Mexico and Central America

ALJOS FARJON

JORGE A. PEREZ DE LA ROSA

& BRIAN T. STYLES

WITH ILLUSTRATIONS BY

ROSEMARY WISE

Published by
The Royal Botanic Gardens, Kew
produced in association with the Oxford Forestry Institute, University of Oxford
1997

First published 1997

ISBN 1 900347 36 9

Cover design by Jeff Eden, page make-up by Media Resources,
Information Services Department, Royal Botanic Gardens, Kew

Production Editor: Suzy Dickerson

Printed in The European Union
by
Continental Printing, Belgium.

Contents

Introduction · · · 1

How to use this guide · · · 3

Distribution · · · 5

Identification · · · 8

Glossary · · · 18

Field Keys · · · 22

Species accounts · · · 36

Literature · · · 122

Index to species · · · 123

Species maps · · · 126

Introduction

Pines (genus *Pinus*, family Pinaceae) are major timber trees in many countries, either as a natural forest resource or in plantation forestry. There are over one hundred species commonly recognized by taxonomists, all of which are native to countries in the northern hemisphere. North America is especially rich with a total of 65 species, of these 38 occur in North America north of Mexico (according to Flora of North America, Vol. 2, 1993) and 43 occur in Mexico and Central America according to the Flora Neotropica monograph (Farjon & Styles, 1997). Several species have distributions which cross the border between the U.S.A. and Mexico. No species occur in Canada that do not also occur in the U.S.A. and similarly, all 9 species found in Central America are also native in Mexico. From this tally one can conclude that Mexico is the country with more species of pines than any other.

This diversity is even greater when one takes into account the numerous varieties (sometimes subspecies) botanists have recognized, some of which have been elevated to full species by others. There is still disagreement among specialists about the exact number of species, especially in Mexico. Perhaps the Flora Neotropica monograph, on which this field guide is based, will bring more stabilization, yet new species are still being described. Several species have highly variable characters, others are more constant, and only a thorough understanding of the entire genus suffices to evaluate these new species. This guide aims at identification of well established species, but addresses variation and problems of species recognition where they occur.

Pines are ecologically a major component of the upland forests of Mexico; they are also abundant or dominant in parts of Guatemala, Belize, Honduras, El Salvador and Nicaragua. In the Central American countries, *Pinus caribaea* var. *hondurensis* occupies the coastal lowlands of the Caribbean Sea, where it forms "pine savannas" over large areas; in Mexico it is only known from a small population in Quintana Roo. The most widespread types of forest in which pines are a major component are pine forests and pine-oak forests. Naturally, there are gradients where pine-oak forest merges into oak forest; increasingly this may be caused by selective cutting of pines. There are considerable differences between pine forests brought about by climatic conditions, since different species are adapted to different climatic conditions. At the driest extreme are the pinyon pines and some of their relatives, bordering semi-desert and desert vegetation. In contrast, some pines occur in extremely wet and cool high altitude forests, often with other conifers such as firs (*Abies*), Douglas fir (*Pseudotsuga*) and Mexican cedar (*Cupressus lusitanica*). Other species occupy intermediate zones and some are only scattered components of other types of forest.

Mexican, and to some extent Guatemalan pine forests are notable for their mixtures of several species growing on the same sites or in close proximity. This diversity diminishes in places where conditions are less favourable to the growth of pines, so that we find fewer species at low and very high altitudes and in dry and very wet locations.

The various uses of pines in Mexico range from timber to firewood and resin to edible seeds. Production of pine logs in Mexico reached a staggering 7.5 million cubic metres in 1986 (Carvajal & McVaugh, 1992). Honduras produces more pine timber than any of the other Central American countries and a substantial section of its economy is based on it. Wood production for lumber is most important, followed by pulped wood for kraft paper and cardboard and local uses for firewood, fence posts etc. Its importance lies in the fact that the xylem of most "hard" pines produces long fibres, giving the final product extra strength. "Soft" pines, less abundant, are sought for their less resinous, evenly grained wood. Resin tapping is commercially important in many regions of Mexico and Central America and several species are being exploited, mostly *P. oocarpa*, *P. montezumae*, *P. teocote* and *P. pseudostrobus*. Pine resin forms the basis for the turpentine industry, which is a major source of revenue in several Mexican states. Edible seeds are provided by the pinyon pines, in particular the widespread species *P. cembroides*, which is common in North Mexico. The seeds of these species are relatively large but few per cone and fall with these to the ground where they are eaten by birds and squirrels. Harvest takes place prior to cone fall, the wingless seeds are marketed locally and exported to major cities in Mexico. Crops vary considerably from year to year over large areas and this type of use can be considered a cottage industry of only local significance.

The forests of Mexico and Central America are under threat of overexploitation, clearance for other forms of land use, overgrazing, increase in frequency of fires and other pressures that generally coincide with human population growth. Many large areas have been depleted of their tallest trees, leaving only secondary growth forest of poor quality in a state of permanent disturbance. Along with such depletion genetic diversity has decreased or is threatened to diminish as regrowth of a mature forest with a variety of species more or less in balance with the environment is not allowed to occur. In Mexico, 10 species of pine are listed by IUCN as threatened with extinction in the foreseeable future if no protective measures are taken (Farjon et al., 1993). Forest conservation should, in our view, give priority to this genetic diversity as it is the best buffer against diseases and predators, climatic disturbances or change and provides options for future and as yet unknown uses. Knowledge of this diversity begins with identification of species and varieties. It is therefore important to identify the species in relatively undisturbed forests prior to exploitation if the aim is reforestation that is both successful and ecologically sound.

Identification in the field of the various species of pines is therefore important to forestry, land-use management and nature conservation. In a difficult group of closely related species of trees which often occur together, a tool should be provided to enable the non-specialist to identify them in the field. This field guide wants to provide such a tool. Moreover, as it is based on a comprehensive taxonomic revision of the pines in the same region, more detailed information can be found in that work. Whenever this was considered appropriate, we have referred to the Flora Neotropica monograph (Farjon & Styles, 1997) that forms the base of reference to this field guide to the pines of Mexico and Central America.

How to use this guide

This field guide treats all the species, subspecies and varieties of pines (genus *Pinus*) that occur naturally in Mexico and Central America. The small distribution maps given with each species likewise only deal with their natural ranges. Relatively few species of pine not native to the region seem to have been planted on any significant scale, but because no information was available in sufficient detail to even list them comprehensively, they have been omitted altogether. Consequently, planted trees can only be identified with this guide if they happen to belong to a species which is also indigenous to this region.

Each species is treated under a number, which is used as a reference throughout the guide. This number is the sequence number used in the Flora Neotropica monograph where species are grouped together according to their (tentative) relationships. The same sequence has been followed in this guide, with the omission of species occurring on the Caribbean Islands. The advantages are that similar species are grouped together for easy comparison and that the user may learn something about these relationships. It also emphasizes the fact that the field guide follows the taxonomy of the monograph. Finding a species by a known name is still easy using the index.

The most important part of any field guide is its determination keys, as their use should lead to identification of the tree at hand. Descriptions and illustrations serve as verification tools for the identification. It is therefore emphasized that use of the keys has preference above any other approaches, such as looking at the illustrations or browsing through the descriptions. Many pines are look-a-likes and one may easily miss the diagnostic characters mentioned in the keys that separate them. Yet, for convenience, a more casual approach has been provided in grouping pines by needle-number and cone size. This limits the number of species one has to choose from and may even lead to a correct identification. Careful reading of the full description and comparison with the illustration is absolutely necessary to verify any identification achieved by this method.

There are two sets of keys: regional and by morphological groups. The regions are shown on map 1 and circumscribed at the head of each key. They usually overlap slightly, but in most cases one can be reasonably certain that no other species occur in the area than the ones mentioned. Many users only working with the pines within one of the regions will find these keys most convenient. The second set of keys covers all species treated in this guide, regardless of where they occur. Here one has to go first through the key to the groups and then proceed with the group key chosen. The workings of the keys are explained at the beginning of the chapter Field Keys.

Important morphological identification characters are discussed in the chapter Identification. Here some terminology is explained, but a glossary included in the guide provides concise definitions of most botanical terms used in the guide. An attempt has been made to keep technical terms at a minimum, but they cannot be avoided altogether unless replaced by lengthy circumscriptions. Users are advised to read the chapter on identification before starting to identify pines as this will greatly assist in interpreting the keys and descriptions correctly.

Characters used in the descriptions have been arranged in an orderly manner and the character states mentioned for each species are worded similarly. This makes cross-reference easy, and especially since similar species tend to be grouped together it is advisable to compare a few nearby descriptions of similar species. The illustration shows good examples of the main characters, but it must be borne in mind that many species are variable. Description and illustration are therefore complementary. The description is restricted to the species in a strict sense (excluding subspecies or varieties when these are recognized). Information on habitat, altitudinal range and distribution (with a small map) pertains to the species including its subspecies or varieties. The latter are mentioned in a last section, where differences with the "species as to its type" are indicated.

Scientific names of species (binomials) consist of a generic name, written with a capital first letter (*Pinus*) and a specific epithet, always written in lower case, both in italics. Subspecies and varieties of species are named with an added epithet and an abbreviated indication of rank. Names of (validating) authors have been added to each taxon in the descriptions, but not in the main text or in the keys. Scientific names are forever changing, primarily due to changing taxonomy, but in part also for purely nomenclatural reasons. The latter should lead to stabilization eventually. For these reasons, a number of names in this guide, since they follow the latest critical revision of Mexican and Central American pines, differ from those used earlier by Martínez (1948) and later publications such as Perry (1991) that were largely based on his work. For details on why such names had to be changed one is referred to Flora Neotropica Monograph 75. (Farjon & Styles, 1997) The different names used by Martínez and Perry are mentioned as synonyms, to enable cross-reference with these often used books as well as other literature. Less used and "forgotten" synonyms have been omitted in this guide, but they are listed in the monograph.

Local or vernacular names are only given in Spanish. They were mainly taken from various literature sources. The diffculty with interpreting these names is that local names may apply to more than one species, or that one species is known by various names in different regions. Some species are not distinguished and are simply called "pine" or an equivalent. There are undoubtedly more names for several species than we have been able to find, and some interpretations may be incorrect. As far as we know, no attempt at standardization has been undertaken. It is therefore advised to memorize and refer to the scientific names of pines preferably.

Much has been written on the pines of Mexico and Central America. Only those publications that are of importance for identification or related matters and are recent or still widely used are listed in the references. For a more comprehensive bibliography one is referred to the monograph, but this publication does not cover papers on forestry aspects.

Distribution

The pines of Mexico and Central America occur throughout the highlands of Mexico and in the highlands of Guatemala, in Belize, Honduras, El Salvador and Nicaragua. In northern Mexico, there are several species with distributions across the border with the U.S.A., some with their main ranges in Mexico, others with a primarily northern distribution. Especially in Baja California Norte, the pines found there form southern extensions or outlying populations of wider distributions in the U.S.A., most of them are Californian pines. All pines occurring in Central America are also found in Mexico, although some species have their main distribution outside Mexico, these are Mesoamerican pines. Several species have very wide distributions, ranging from northern or central Mexico well into Central America, or across a large part of Mexico. A larger number of species have a more restricted range, more or less contiguous or often broken up into several disjunct populations. Finally, there are a number of species and subspecies with small ranges, sometimes limited to one or a few locations, these are local endemics.

Pines occur from the coastal lowlands bordering the Caribbean Sea up to the tree line on the highest volcanos of Mexico, around 4000 m altitude. Very few species are restricted to the lowlands: *P. caribaea* var. *hondurensis* in extensive "pine savannas" of Central America and *P. muricata* in a few localities along the coast of the Pacific Ocean in Baja California Norte. The vast majority of the pines are montane to high montane, while some have very wide altitudinal ranges and occur from the foothills to 3000 m. Species with a wide altitudinal range are also widespread, and often occur at higher altitudes towards the NW of their distributions. In table I the pines of Mexico and Central America have been grouped according to six regions and four altitudinal categories. A species can occur in more than one of each.

The geographical regions are: **I.** Californian; **II.** NW Mexican; **III.** W Mexican; **IV.** NE & E Mexican; **V.** Central & S Mexican; **VI.** Mesoamerican.
The altitudinal categories are: **1.** lowland (1–300(–700) m); **2.** foothills ((100–)300–1200 m); **3.** montane ((700–)1000–2600(–2800) m); **4.** high montane (2000–)2500–4000(–4300) m).

In the chapter Field Keys, a separate key will be presented for the species in each of the six geographical regions. This should limit the number of species one has to consider substantially; however, the possibility that a species occurs outside the area, of which the boundary cannot be precisely drawn, cannot be ruled out entirely. The six regions are indicated on map 1.

TABLE 1

4		*P. cembroides* var. *bicolor* *P. flexilis* var. *reflexa* *P. strobiformis* *P. teocote*	*P. hartwegii* *P. montezumae* *P. pseudostrobus* + var. *apulcensis* *P. strobiformis* *P. teocote* *P. montezumae* *P. pseudostrobus* *P. teocote* *P. strobiformis*	*P. ayacahuite* + var. *veitchii* *P. culminicola* *P. flexilis* var. *reflexa* *P. hartwegii*	*P. ayacahuite* *P. hartwegii* *P. montezumae* *P. pseudostrobus* + var. *apulcensis* *P. teocote*	*P. ayacahuite* *P. hartwegii* *P. montezumae* *P. pseudostrobus* + all varieties
3	*P. contorta* var. *murrayana* *P. coulteri* *P. jeffreyi* *P. lambertiana* *P. monophylla* *P. quadrifolia* *P. radiata* var. *binata*	*P. arizonica* + var. *cooperi* *P. cembroides* *P. devoniana* *P. engelmannii* *P. leiophylla* var. *chihuahuana* *P. leiophylla* var. *leiophylla* *P. oocarpa* *P. teocote*	*P. cembroides* subsp. *lagunae* *P. devoniana* *P. douglasiana* *P. durangensis* *P. herrerae* *P. jaliscana* *P. leiophylla* var. *chihuahuana* *P. leiophylla* var. *leiophylla* *P. maximartinezii* *P. maximinoi* *P. montezumae* *P. oocarpa* *P. praetermissa* *P. pseudostrobus* + var. *apulcensis* *P. rzedowskii* *P. teocote*	*P. arizonica* var. *stormiae* *P. cembroides* + subsp. *orizabensis* *P. devoniana* *P. engelmannii* *P. greggii* *P. leiophylla* var. *leiophylla* *P. montezumae* *P. nelsonii* *P. patula* + variety *P. pinceana* *P. pseudostrobus* *P. remota* *P. teocote*	*P. cembroides* *P. devoniana* *P. douglasiana* *P. herrerae* *P. lawsonii* *P. leiophylla* var. *leiophylla* *P. maximinoi* *P. montezumae* + var. *gordoniana* *P. oocarpa* + var. *P. patula* + var. *P. pringlei* *P. pseudostrobus* + var. *apulcensis* *P. strobus* var. *chiapensis* *P. tecunumanii* *P. teocote*	*P. devoniana* *P. tecunumanii* *P. maximinoi* *P. montezumae* *P. oocarpa* *P. pseudostrobus* + all varieties *P. strobus* var. *chiapensis*
2	*P. attenuata*				*P. devoniana* *P. maximinoi* *P. oocarpa*	*P. devoniana* *P. maximinoi* *P. oocarpa* *P. tecunumanii*
1	*P. muricata*					*P. caribaea* var. *hondurensis*
	I	**II**	**III**	**IV**	**V**	**VI**

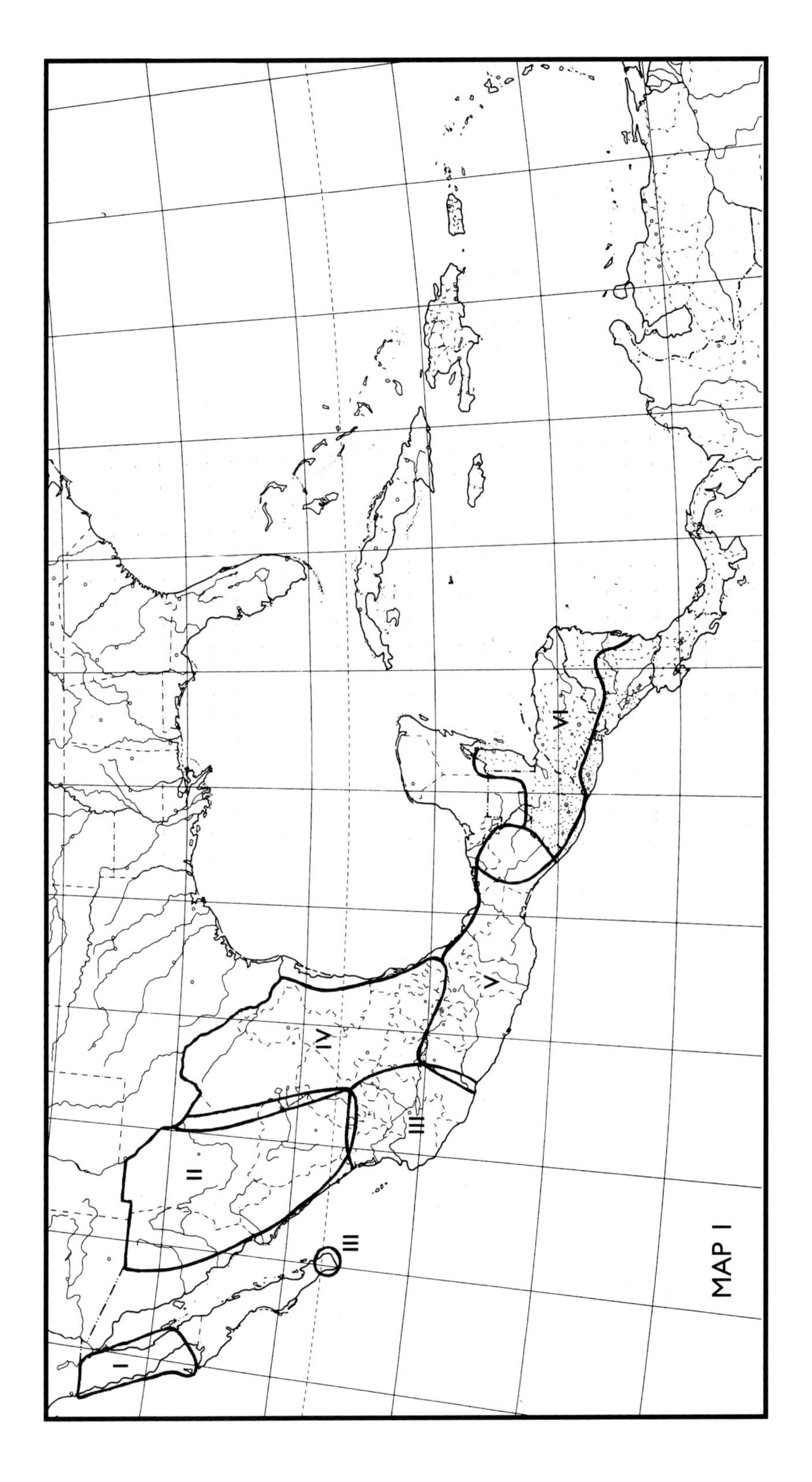
MAP I
I
II
III
III
IV
V
VI

Identification

Most pines are at first sight very similar and many characters by which they might be distinguished are extremely variable. This difficulty makes certain identification problematic, especially for the beginner. It also means that few pines can be identified by a single character. A combination of characters, together uniquely true for only one species, has to be evaluated in order to identify the pines of Mexico and Central America with certainty. Usually, both foliage and cones are needed to recognize a sufficient number of characters, but (mature) cones may not always be available. Not all pines grow everywhere, although some species are extremely widespread, and so one might be tempted to consider distribution (geographical and altitudinal) as a "character" for identification. However, for several species our knowledge of their ranges may still be incomplete, or trees may have been planted or seeded outside their natural range. Although we shall treat some regions and their species separately in this guide, morphological characters should have preference for identification purposes.

There are several methods available to a field worker to identify a tree. Each of these may by itself lead to a satisfactory result, but often a combination is advisable.

1. Collect material, preferably foliage and cones, and take it to a botanical institution for identification. Comparison with identified herbarium specimens and/or examination by a botanist who has a knowledge of pines will usually identify the species.

2. Look through the illustrations of this guide until one is found that seems to fit the tree in question, then check the accompanying description carefully. Usually a reference to related or similar species is made, these descriptions should be checked as well.

3. Ask the vernacular name from a local villager, look up this name in the guide, then read the descriptions under these names carefully. For pines, this is a less satisfactory method since few species are distinguished by the local people. Different species may be called the same, or a species may be known by different names.

4. Use the identification keys (the botanical and preferred method). How to use the keys will be explained further down, suffice it to say here that with some practice they will enable the field worker to identify any species of pine that occurs naturally in Mexico and Central America. The keys lead to a single species, but nevertheless the description should be read carefully as a check against errors made when reading through the keys.

When an identification is made which seems to be wrong according the the information given for distribution and/or altitude, a further check in a herbarium or by a specialist is desirable. Also, one should consult the monograph of pines in Flora Neotropica (Farjon & Styles, 1997), providing more detailed descriptions and other information which forms the basis from which this guide has been compiled. To do this easily, the numbering of species in this guide is kept identical with that in the monograph. When identification seems certain and the tree is found to grow in the wild, it could appear that a new locality has been discovered. In such a case it is desirable to deposit at least one specimen in one or more institutional herbaria. It is by these proven records that reliable distribution maps of species are compiled and this information is in turn important for forestry and conservation policies.

Even with all the field characters available, some pines may turn out to be impossible to identify with certainty. Theoretically, there would be the possibility that a new species has been discovered. However, the likelihood of this becomes more remote with the steady increase in collecting throughout Mexico and Central America during the past decades. Although most of the reported hybridization between Mexican pines is putative rather than verified, it is likely that several species do hybridize. Especially F^1 trees might combine characters of either parents in a way not encountered in any of the species. Another, and possibly related, feature is the continuous nature of many characters, especially those of size or number, found in closely related species. Because of this, it can become difficult to identify a tree unambiguously. Traditionally, such species have been treated in "species groups" or "species aggregates", but delimitation of these groups appears often as problematic as that of the constituent species. Field characters may turn out to be inadequate to identify the species, and laboratory methods may then be the only further option.

Identification characters

The characters used in the keys are those given in the descriptions and illustrations of the species. The most important of these involve the foliage (twigs and needles) and the cones. In addition, habit and bark features are sometimes mentioned, but they are, with exceptions, less reliable. Several characters separate groups of species, rather than single species, most notably the two subgenera recognized in the genus. Such characters are important and used in the keys to narrow the choice of species, but they are not repeated in the descriptions of the species. Although terminology has been kept to a minumum, some technical terms are necessary for accurate description and need to be learned. They will be explained in the descriptions of the characters and in the glossary.

Habit, trunk

With the habit is meant the overall appearance of the plant. In this guide a mere distinction is made between shrub and tree and no crown shape is mentioned in most cases. This is omitted because it is highly dependent on environmental factors, including the proximity of other trees. A shrub is here defined as a low branching or multi-stemmed pine not taller than 5–6 m and with a crown usually as broad as it is high. Heights of trees are given as maximum heights, with a range according to performance on poor or good growing sites. Similarly, the girth is usually a maximum, given as diameter at breast height (d.b.h.) and should be read as an indication of measurement among large trees. The trunk is the main stem from the ground up to where it is "lost" in branching. All young pines and mature trees of some species are monopodial to the top, which means that branches are in lateral whorls from a main stem. However, this regularity is usually lost in the crown of older trees. The shape of the trunk is naturally subject to environmental conditions.

Bark

The bark of pines, as commonly used for identification of species, is the outer bark or dead tissue on the trunk. It usually breaks into small or large plates. When these plates remain attached to newer bark layers, the outer bark becomes a thick, rigid covering and because the trunk continues to increase its girth fissures appear. Plates and fissures often have a characteristic pattern, but this pattern is variable and rarely unique for a single species. In a few species it is quite distinct and can be used to identify the tree, in most cases it is only useful in combination with other characters. The bark of young trees and branches is usually less characteristic, so it is not used in this guide; bark of large trunks at about eye-level should be used for comparison.

Foliage twigs

The foliage of pines, although evergreen, persists for a limited time, usually only a few years. Some high altitude (subalpine) species retain their needles longer. The twigs bearing needles also feature a number of other characters, which are lost later when bark is formed. In most pines a new length of shoot is formed each season, terminating in a "winter bud", but some species may have two or more bursts of growth in a single growing season, especially on leading shoots. These are termed multinodal and can be recognized by interruptions (nodes) in a single year's growth. Sometimes more than one whorl of cones appears as well. The young shoots are either green or orange-brown, in some species they are distinctly glaucous (with a bluish bloom which soon disappears). The needle fascicles stand on slight elevations (fascicle bases or pulvini), which are long, decurrent, forming ridges, or small and short. This character divides the two main groups: subgenus *Pinus* with ridged shoots and subgenus *Strobus* without ridges. The ridges are most prominent on thick, vigorous leading shoots.

Needles

Pines have four types of leaves (two types in the mature plant) but for this guide we are only concerned with the most conspicuous and distinct type: the needles. The needles of pines grow in fascicles from dwarf shoots (not longer than a few mm) and fall bundled together. At their base a number of scales forms a sheath, which may be deciduous or persistent. The characters of the fascicle sheaths are important to identify groups rather than individual species, but in combination with other features they can sometimes help to identify a species as well. The number of needles per fascicle is one of the most widely used and quoted characters of pines. Yet, with some exceptions, it is a variable character. The number of needles varies on the same tree (branch), and sometimes among trees of the same species. It is therefore important to count a sufficient number of not too old fascicles on various branches of a tree. Ranges are given when variation occurs in a species; the more prevalent numbers are given surrounded by less often found numbers in parentheses. It is important to note that this often skewed range is more significant than an average number with minimum and maximum, as is sometimes done. The length and width of the needles are given in the same manner. The consistency of the needles, often a function of the ratio length/width, is important for the "habit" of the foliage: straight or curved and rigid (spreading or upright) are quite distinct from drooping or pendulous and lax and often determine the appearance of a tree from a distance. Unfortunately, there are many grades between the two extremes and some species may be highly variable. The colour of the needles, usually some shade of green, is likewise variable and is only given in this guide when it is distinct and constant. With a hand lens (×10) rows of stomata can be seen running lengthwise either on all sides of the needles or only on the two inner sides. Their number, though variable, is to some extent related to the width of the needles and therefore not always reliable. Again, information on the position of the stomata is limited to those species where it is an informative character. The margins of most pine needles are finely serrulate (saw-toothed) but some are only weakly serrulate (with the tiny teeth far apart) or entire. This character is only mentioned where it helps to identify the species. Finally, there are anatomical characters, such as the position of resin ducts as seen in cross section, which are diagnostic especially in combination. However, for a field guide they are less useful as they can only be seen clearly with the help of a light microscope. Information is given in some detail in the monograph (Farjon & Styles, 1997), it may be needed in those cases where laboratory work on difficult specimens is considered necessary.

Cones

The cones (we only consider the seed cones, not the pollen cones) remain the most distinctive features for the identification (and classification) of the pines. In no other genus of conifers are they as diverse as in *Pinus*, although between several species there is hardly any difference. The cone crop is not abundant every year, but usually some old cones persist or can be found under the tree. In the latter case care must be taken not to collect something that might belong to another tree: where apparently different pines grow together only material attached to the tree should be used. Important are the placement of the cones on the branch, their position at maturity, their persistence (from falling soon after seed dispersal with the peduncle attached, or falling later leaving some scales on the branch, to not falling at all), the length and shape of the peduncle or stalk, the time it takes for the scales to open, the shape of the cones both with closed and opened scales, and their size. Colour is sometimes distinct, as is the amount of resin exuded, but most cones are a dull brown, darker "inside" (the scales proper) than on the exposed apophyses.

Cone scales

The number of cone scales is often independent of the size of the cone and although variable, is given in the descriptions. By counting the number of scales in a spiral from base to apex and determining the number of spirals it is usually quite easy to estimate the number of scales. The scales are attached to an axis (usually firmly, but in some species rather weakly) and are variously thin or thick and rigid. On the inside, often two lighter coloured marks of the seed wings are visible. In some species, the wing is rudimentary and the seeds are embedded in cup-like depressions. The most prominent feature is the apical part of the scale, consisting of an apophysis and an umbo. The apophysis is the thickened part which is exposed and the only part of the scale visible when the cone is full grown but still closed. The umbo is a smaller, central or terminal portion which represents the initial growth of the two-phased (rarely three-phased) development of pine cones. It is the apophysis which is the most variable between or within species. Flat, raised or even pyramidal, conical or curved, often transversely keeled and variable on different parts of the cone, with the strongest development often on one side, the apophyses together determine much of the characteristic features of the cone. The umbo, terminal in some species but mostly dorsal (in the centre of a transverse keel or of ridges across the apophysis) ends in a small spine or prickle, which in most species disappears in mature cones but in some species, where the spine is larger, persists.

Seeds

On each cone scale, two seeds develop. A seed wing develops from the tissue on the inside of the cone scale (or it remains rudimentary) and is variously attached to the seed. When the wing is easily removed (only held to the seed by two appendages) it is articulate; when it is firmly attached and usually breaks off it is adnate. Wingless seeds do not retain any part of the wing (which is rudimentary) and, although easily removed, do not fall from the scale because they are embedded in a cup-like depression. The sizes of seeds and wings are usually correlated to the sizes of the scales (and cones), but the wingless seeds are large relative to the size of the scales. In some cases the relative length of the seed wing is an important character. All winged seeds are slightly flattened. Colours of seeds and wings vary, usually the seeds are darker than the wings and often there are dark spots on the seeds and dark stripes on the wings. In the wingless seeds, there are differences in the thickness of the seed coat (integument) which can be important to identify the species.

Categories of characters

Certain categories of characters distinguish a limited number of species; it can be useful to check if the tree at hand has characters in one or more of these categories to either limit the number of possibilities or to identify the tree. Species with such less common characters are listed under each category, with a brief description of the character found in each species.

1. Number of needles per fascicle

1.1 One (rarely two) needle per fascicle

46. *Pinus monophylla*	Needles in fascicles of 1, rarely 2, curved, (2–) 2.5–6 cm long, 1.2–2.2(2.5) mm wide, round, acute, often glaucous. Baja California.

1.2 Two needles per fascicle

3. *Pinus contorta* var. *murrayana*	Needles in fascicles of 2, 4–7 cm long, 1.2–2.0 mm wide, margins entire or weakly serrulate. Baja California.
29. *Pinus muricata*	Needles in fascicles of 2, (7–)10–14(–16) cm long, 1.3–2.0 mm wide, margins serrulate. Baja California mainland.
30. *Pinus radiata* var. *binata*	Needles in fascicles of 2, sometimes 3 on leading shoots, 8–15 cm long, 1.1–1.6 mm wide, margins serrulate. Baja California islands.

1.3 Two or/and three needles per fascicle

43a1. *Pinus cembroides* var. *cembroides*	Needles in fascicles of 2–3, (2–)3–5(–6.5) cm long, (0.6–)0.7–1.0 mm wide, stomata on all sides, margins entire. Fascicle sheaths recoiling to form a rosette, deciduous.
9. *Pinus ponderosa*	Needles in fascicles of (2–)3, (10–)15–25(–27) cm long, 1.3–1.6 mm wide, margins serrulate. Fascicles sheath persistent.
45. *Pinus remota*	Needles in fascicles of 2(–3), (2–)3–4.5(–5.5) cm long, 0.8–1.1 mm wide, stomata on all sides, margins entire. Fascicle sheaths soon deciduous, not recoiling to form a rosette.

1.4 Three needles per fascicle

31. *Pinus attenuata*	Needles in fascicles of 3, rarely 2, (8–)10–12 (–14) cm long, 1.0–1.5 mm wide, rigid, spreading. Baja California.
33. *Pinus coulteri*	Needles in fascicles of 3, 15–25(–30) cm long, 1.9–2.2 mm wide, very thick and rigid, spreading. Shoots multinodal, thick. Baja California.

32. *Pinus greggii*
Needles in fascicles of 3, (7–)9–13(–15) cm long, 1.0–1.2 mm wide, rigid, spreading.

4. *Pinus herrerae*
Needles in fascicles of 3, (10–)15–20 cm long, 0.7–0.9 mm wide, slender, lax, drooping or spreading.

11. *Pinus jeffreyi*
Needles in fascicles of 3, rarely 2, (12–)15–22 (–25) cm long, 1.5–1.9(–2) mm wide, thick, rigid, spreading. Shoots uninodal, slender. Baja California.

19. *Pinus lumholtzii*
Needles in fascicles of 3, rarely 2 or 4, (15–) 20–30(–40+) cm long, (1–)1.2–1.5 mm wide, lax, pendulous. Fascicle sheaths deciduous.

41. *Pinus nelsonii*
Needles in fascicles of 3, rarely 4, 4–8(–10) cm long, 0.7–0.8 mm wide, connate (seemingly single), margins serrulate, spreading.

20b. *Pinus oocarpa* var. *trifoliata*
Needles in fascicles of 3, rarely 4, (11–)14–17 (–20?) cm long, 1.2–1.6 mm wide, straight, rigid, spreading.

42. *Pinus pinceana*
Needles in fascicles of 3, rarely 4, 5–12(–14) cm long, 0.8–1.2 mm wide, margins entire, spreading. Fascicle sheaths deciduous.

1.5 Three or/and four needles per fascicle

10c. *Pinus arizonica* var. *stormiae*
Needles in fascicles of 3–4, rarely 5, 14–25 cm long, 1.4–1.8 mm wide, rigid, usually curved and twisted, spreading. NE Mexico.

5c. *Pinus caribaea* var. *hondurensis*
Needles in fascicles of 3, rarely 2, 4, very rarely 5, (12–)16–28 cm long, (1.2–)1.4–1.8 mm wide, straight, spreading. Central America.

43c. *Pinus cembroides* subsp. *orizabensis*
Needles in fascicles of 3–4, rarely 2 or 5, (2–) 3–5(–6) cm long, 0.7–1.1 mm wide. Fascicle sheaths recoiling and deciduous.

12. *Pinus engelmannii*
Needles in fascicles of (2–)3(–4), rarely 5, (18–) 20–35 cm long, 1.5–2.0 mm wide, very thick, rigid, spreading.

1b. *Pinus leiophylla* var. *chihuahuana*
Needles in fascicles of (2–)3(–4), rarely 5, (4–) 6–12(–14) cm long, 0.9–1.3(–1.5) mm wide, rigid, spreading. Fascicle sheaths deciduous.

27. *Pinus pringlei*
Needles in fascicles of 3(–4), (15–)18–25(–30) cm long, 1–1.5(–1.7) mm wide, rigid, straight, spreading. S Mexico.

1.6 Five needles per fascicle, sheath persistent

16. *Pinus devoniana*
Needles in fascicles of 5, rarely 4 or 6, (17–) 25–40(–45) cm long, 1.1–1.6 mm wide, slender, spreading. Fascicle sheaths very long, up to 40 mm, often resinous.

17. *Pinus douglasiana*
Needles in fascicles of 5, rarely 4 or 6, 22–35 cm long, 0.7–1.2 mm wide, lax, drooping.

18. *Pinus maximinoi*
Needles in fascicles of 5, rarely 4 or 6, 20–35 cm long, 0.6–1.0(–1.1) mm wide, very thin, lax, drooping or pendulous.

14. *Pinus pseudostrobus*
Needles in fascicles of 5, rarely 4 or 6, (18–) 20–30(–35) cm long, 0.8–1.3 mm wide, slender, lax, spreading, drooping or pendulous.

1.7 Five needles per fascicle, sheath deciduous

34. *Pinus ayacahuite*
Needles in fascicles of 5, very rarely 6, (8–) 10–15(–18) cm long, 0.7–1.0 mm wide, margins (weakly) serrulate, stomata on two inner sides only. Central Mexico to Central America.

44. *Pinus culminicola*
Needles in fascicles of 5, very rarely 4 or 6, 3–5 cm long, 0.9–1.3 mm wide, margins very weakly serrulate or entire, stomata on two inner sides only. Fascicle sheaths recoiling. NE Mexico.

36. *Pinus flexilis* var. *reflexa*
Needles in fascicles of 5, (5–)6–9 cm long, 0.8–1.2 mm wide, margins very weakly serrulate or entire, stomata mostly on two inner sides, a few also on the outer side.

35. *Pinus lambertiana*
Needles in fascicles of 5, (3.5–)4–8(–10) cm long, 0.8–1.5 mm wide, margins very weakly serrulate or nearly entire, stomata on all sides. Baja California.

40. *Pinus maximartinezii*
Needles in fascicles of 5, very rarely 3 or 4, 7–11(–13) cm long, 0.5–0.7 mm wide, margins entire, stomata on two inner sides only. Fascicle sheaths recoiling. Mexico: Zacatecas.

37. *Pinus strobiformis*
Needles in fascicles of 5, very rarely 6, (5–) 7–11(–12) cm long, (0.6–)0.8–1.1(–1.2) mm wide, margins weakly serrulate or nearly entire, stomata usually on two inner sides only. N Mexico.

38. *Pinus strobus* var. *chiapensis*
Needles in fascicles of 5, (5–6–12(–13) cm long, 0.6–0.8(–1.0) mm wide, margins weakly serrulate, stomata on two inner sides only. S Mexico, W Guatemala.

All other pines in Mexico and Central America have such variation in needle numbers, that they cannot be reliably assigned to any of these categories. This does not mean that some trees of these species cannot have a fairly constant number of needles (such as is regularly the case with *Pinus montezumae* having predominantly 5 needles), but it means that other trees of the same species will be more variable.

2. Maximum and minimum length of mature cones

2.1 Cones shorter than ca. 6 cm

43. *Pinus cembroides*	Cones (2–)3–5(–7.5) × 3–6(-7) cm when open; scales 25–40(–50), with deep depressions holding large, wingless seeds. Needles in fascicles of (2–)3–4, rarely 5.
3. *Pinus contorta* var. *murrayana*	Cones (3–)4–5.5 × 3–4 cm when open; umbos of scales with a prominent, persistent prickle. Needles in fascicles of 2.
44. *Pinus culminicola*	Cones 3–4.5 × 3–5 cm when open, scales 45–60, with deep depressions holding large, wingless seeds. Needles in fascicles of 5, very rarely 4 or 6.
4. *Pinus herrerae*	Cones (2–)3–3.5(–4) × 2–3.5 cm when open; umbos with a minute, deciduous prickle. Needles in fascicles of 3.
19. *Pinus lumholtzii*	Cones (3–)3.5–5.5(–7) × (2.5–)3–4.5 cm when open; umbos flat, with a minute, deciduous prickle. Needles in fascicles of 3, rarely 2 or 4, pendulous; fascicle sheaths long, deciduous.
46. *Pinus monophylla*	Cones 4–6 × 4.5–7 cm when open; scales 30–50, with deep depressions holding large, wingless seeds. Needles in fascicles of 1, rarely 2. Baja California.
21. *Pinus praetermissa*	Cones (4–)5–6.5(–7) × (5–)6–8 cm when open, broadly ovoid and smooth when closed; scales near the base often fall before the cone. Peduncle slender.
47. *Pinus quadrifolia*	Cones 4–6 × 4.5–7 cm when open; scales 30–50, with deep depressions holding large, wingless seeds. Needles in fascicles of (3–)4(–5), rarely 2 or 6. Baja California.
45. *Pinus remota*	Cones (2–)2.5–4 × 3–6 cm when open; scales 25–35, with deep depressions holding large, wingless seeds with very thin seed coat (integument). Needles in fascicles of 2(–3). NE Mexico.
24. *Pinus tecunumanii*	Cones (3.5–)4–7(–7.5) × (3–)3.5–6 cm when open, ovoid to broadly ovoid when closed; scales persistent. Peduncle stout, strongly curved.
28. *Pinus teocote*	Cones (3–)4–6(–7) × 2.5–5 cm when open, ovoid to ovoid-oblong when closed; scales with a flat umbo. Peduncle very short.

Some other species may have cones shorter than 6 cm (notably *Pinus oocarpa*), but the variation is considerable and there are many trees with larger cones. *Pinus tecunumanii* is also variable but is listed here because most trees have cones shorter than 6 cm.

2.2 Cones longer than ca. 15 cm

34. *Pinus ayacahuite*	Cones (10–)15–40(–50) × 7–15 cm when open; scales thin, flexible, usually recurved or reflexed, with terminal umbos. Seed wings longer than seeds (look for seed wing marks on the cone scales).
33. *Pinus coulteri*	Cones 20–35 × 15–20 cm when open, very heavy, with very large, curved apophyses and umbos. Baja California.
16. *Pinus devoniana*	Cones 15–35 × 8–15 cm when open; scales thick, umbos dorsal on a transverse keel of the apophysis.
35. *Pinus lambertiana*	Cones 25–45 × 8–14 cm when open; scales thick, nearly straight, with terminal umbos. Baja California.
40. *Pinus maximartinezii*	Cones (15–)17–25(–27) × 10–15 cm, very heavy; scales thick, with deep depressions holding large, wingless seeds and with very thick apophyses and dorsal umbos. Zacatecas.
37. *Pinus strobiformis*	Cones 12–30(–60) × 7–11 cm when open; scales thin but rigid, usually (strongly) recurved, with terminal umbos. Seed wings shorter than seeds (look for seed wing marks on the cone scales).

There are some other pines with cones occasionally longer than 15 cm (but only a few cm at most), but the great majority of the cones of these species is shorter. The object of the listing on cone sizes is to separate the species with small and large cones for quick identification. It is advisable to check at least the full description of the species chosen from these lists before making a definite identification.

Glossary

In the following brief definitions of botanical terms used in this guide are provided. The definitions given here are only valid within the context of this guide (as pertaining to pines) and could therefore differ from a more general meaning used elsewhere. It is assumed that a basic knowledge of botany is present so that only more specific technical terms are explained here. Illustrations are given to help identify the structures to which several terms refer. Terminology pertaining to microscopic features inside the needles has been omitted; for these the monograph (Farjon & Styles, 1997) should be consulted.

TERM	DEFINITION
abaxial	situated on the side away from the axis (shoot), on pine needles on the outer side when all needles in a fascicle are brought together.
adaxial	situated on the side towards the axis (shoot), on pine needles on the inner side touching the other needles when all in a fascicle are brought together.
adnate	having a firm connection between the seed and the seed wing; the wing breaks off when an attempt is made to pull it free (see Fig. 2J).
amphistomatic	with stomata situated on all sides of the needles, although often there are more stomata on one (or two adjacent) sides.
apophysis	the exposed part of the cone scale when the cone is full-grown but still closed; it is usually thicker and harder than the lower part of the scale (see Fig. 2C–D).
articulate	having a loose (jointed) connection etween the seed and the seed wing; the wing can be pulled from the seed with ease or falls off by itself (see Fig. 2K).
attenuate	with a narrowing shape, tapering to a point.
cataphyll	the small, non-green, thin scale-like leaf at the base of a fascicle; in some species the cataphylls persist longer than the needles, in others they fall earlier (see Fig. 1C, 1F).
connate	having a firm connection of two similar organs (e.g. needles, compare adnate) which retain their morphological and physiological independence.
deciduous	shedding organs (e.g. needles, cones), either at intervals or more or less continuously.
decurrent	gradually disappearing down the shoot to which the structure (leaf base) is tightly attached. The leaf bases form ridges on the shoot (see Fig. 1A).
distal	positioned at the end or top.

epistomatic with stomata all positioned on the adaxial side(s) of the needles.

fascicle the bundle of needles, varying in number from (1–)2–5 (–8) which grows from a dwarf shoot and usually falls in its entirity (see Fig. 1G).

fascicle sheath the thin scales surrounding the basal part of the fascicle, which either form a tight and persistent wrapping or disintegrate when the needles are full grown (see Fig. 1H).

"grass stage" a stage of growth in the seedling in which the stem does not grow and only the root system and needles are expanded; such seedlings look like tufts of grass on the ground.

mucronate abruptly ending in a small pointed appendix.

multinodal a shoot is multinodal when it produces more than one flush of growth in a year, each terminating in a whorl of branches and sometimes cones. A year's growth may be more difficult to distinguish especially on leading shoots, but is usually determinable.

peduncle the stalk of a cone (pedunculate = stalked).

proximal positioned at the base or bottom end.

pubescent (densely) covered with short hairs.

pulvinus the elevated base of the scale leaves (cataphylls), which is decurrent or non-decurrent (see Fig. 1A–B).

serotinous cones remain closed for a considerable time or require heating by fire to open (orig. Latin = late-coming).

serrulate finely serrate = margins with minute saw-teeth pointing forward. The teeth can be remote (a few per mm or less) or dense (see Fig. 1I–K). If they are absent the margins are described to be entire.

umbo (orig. Latin = the boss of a shield); in pine cones it is a prominence on the apophysis and represents the exposed part of the scale at the initial growth phase. It may be blunt or armed with a prickle or spine (see Fig. 2D). Its position is terminal (at the true tip of the apophysis) or dorsal ("on the back").

uninodal a shoot is uninodal when it produces a single flush of growth in a year, terminating in a whorl of "winter buds" which contain the beginnings of next year's growth.

vestigial reduced to a remnant, usually incapable of performing the functions of a fully developed organ (e.g. seed wing; also: rudimentary).

winter bud a terminal or subterminal vegetative bud containing new needle fascicles and the shoot apex; it is covered by scale leaves which remain at the base of the fascicles in many species (see Fig. 1C).

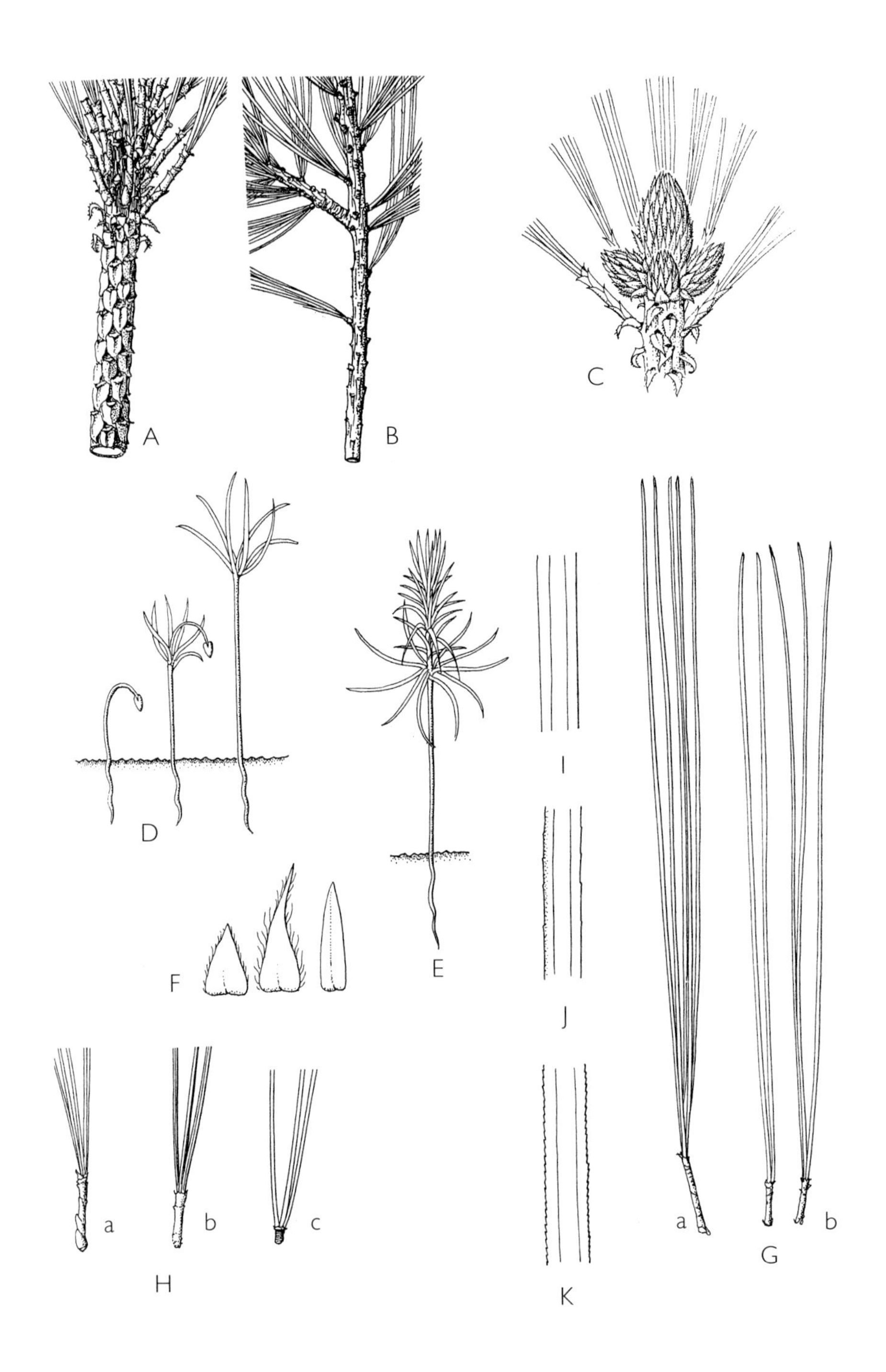

FIG. 1. Morphology of foliage in *Pinus*. **A.** branchlet with decurrent leaf bases, **B.** branchlet with non-decurrent leaf bases, **C.** "winter bud" complex, with terminal and subterminal buds, **D.** seedling with cotyledons, **E.** seedling with primary leaves, **F.** types of cataphylls, **G.** needles in fascicles,, **H.** fascicle sheaths, **I.** entire leaf margins, **J.** remotely serrulate leaf margins, **K.** serrulate leaf margins.

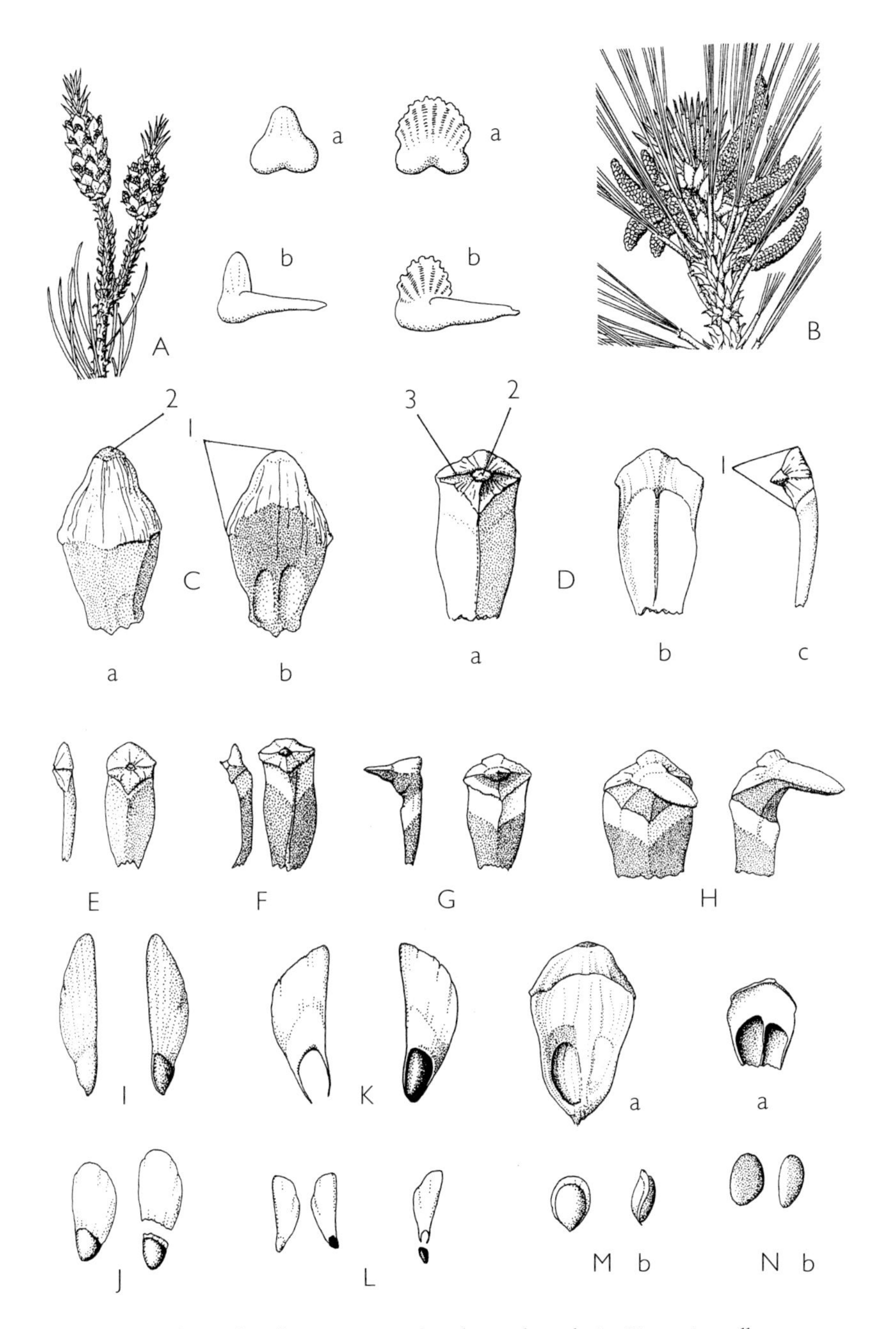

FIG. 2. Morphology of pollen cones, seed scales and seeds in *Pinus*. **A.** pollen cones; microsporophyll type of *P.* subgenus *Strobus*, a = abaxial view, b = lateral view, **B.** idem of *P.* subgenus *Pinus*, **C.** seed scale type with terminal umbo, a = abaxial view, b = adaxial view, 1 = apophysis, 2 = terminal umbo, **D.** seed scale type with dorsal umbo, c = lateral view, 2 = dorsal umbo, 3 = transverse keel, **E–H.** increased development of apophysis and umbo in seed scales, **I–J.** seeds with adnate, effective wings, **K–L.** seeds with articulate, effective wings, **M.** seeds with adnate, ineffective wings, a = seed scale with seed cavity (one seed abortive), b = seeds, **N.** seeds (b) with articulate, vestigial wings remaining on the scale (a).

Field Keys

How to use a key

A key is a method by which one can identify a plant (or any object) by reading brief descriptions and choosing between alternatives. There are various types of keys, but the most commonly used keys in botany are dichotomous keys. In this type of key, each statement has an opposing statement somewhere in the same key, and a choice must be made between the statements of this "couplet" before one can proceed. The couplets can be numbered for easy reference, and the key can be indented to provide a visual aid in recognizing the couplets. The following example will illustrate this principle:

Example key with four plants

1. Plant with fleshy stems lacking leaves, usually with spines concentrated in specific points on all parts · **Cactus**
1. Plant without fleshy stems and spines and at least seasonally with leaves.
    2. Plant with thick woody stem
        3. Plant evergreen, with long, thin needles and woody cones containing the seeds · **Pine**
        3. Plant usually deciduous, with broad, flat leaves and acorns · · **Oak**
    2. Plant with thin herbaceous stem devided by knobby nodes, leaves linear, with a basal sheath surrounding the stem · · · · · · · · · · · **Grass**

It will be seen, that in the first couplet, there are three characters: fleshy stems, leaves and spines. Just how they do or do not apply to the plant in question defines their character state: fleshy stems, leaves and spines are present or absent and spines, if present, are concentrated in clusters and occur on all parts of the plants. The combination of these character states determines whether our plant is a cactus or not: each character by itself is not sufficient. There may be other plants with fleshy stems, without leaves or with spines, but only a cactus combines all three character states.

The opposing statement in couplet No. 1, if true, defines any plant not being a cactus. To proceed if our plant is not a cactus, we must read couplet No. 2 and make a choice. Note that the two statements of this couplet are some distance apart, this is because more than one possibility fits the first statement. If it is true, one proceeds with couplet No. 3 and so on until the correct name of the plant is reached.

In the more complicated keys to the pines that follow, it will be found that the statements often allow for some variation and that it is not always easy to decide. A combination of character states is therefore used to attain more certainty. Often, the variation in size given in the key can or even must be expected to occur in a single tree or at least in a few trees of the same species growing together. It is therefore important to search for this variation in the field. It can also happen, that a feature mentioned in the key is not present at the time (for instance, the cones). It is then often possible to try each of the two choices: it will usually become apparent further on which of the two is wrong. Return in that case (or whenever in doubt) to the last couplet where a certain choice could be made and try the other option. To add more certainty, sometimes the position of resin ducts in the needles is stated; the remark (microscope!) is added because this cannot be checked in the field and requires preparation of cross-sections for a light microscope with ca. 50x magnification.

Because the pines of Mexico and Central America are often quite similar, it is very difficult to key them all out in a single key using only a limited number of field characters. For this reason, groups of keys are constructed, each treating a limited number of species. Groups are defined geographically distinguishing six regions (see chapter on distribution), each with the pines occuring in that region. Additional groups treat pines which share a certain character or combination of characters to the exclusion of other pines. Both groups are artificial, but the second kind tends to group related species more often.

The procedure when using the keys in this field guide is as follows:

1) Start with the key to the morphological groups, or with the key of the geographical group which applies.

2) Decide which statement in the first couplet is more correct.

3) If the correct statement ends
  a) with a group number, then turn to the key of this group;
  b) with a species name, then this name is the identification (but check by reading the description and seeing the illustration).

4) If the first statement is not correct, proceed with the alternative under the same couplet number.

5) Repeat the procedure in the key of the group.

KEYS BY REGIONS I–VI

Key to Species in (insular) Baja California Norte (I)

1. Fascicles with 5 needles; fascicle sheaths deciduous; seed cones ≥25 cm long, pendulous; umbo of seed scale terminal (Fig. 2C)
   . . . . **35.** *P. lambertiana*
1. Fascicles with 1–4 needles, rarely 5; fascicle sheaths persistent or recoiling and deciduous; seed cones small or large, not pendulous; umbo of seed scale dorsal (Fig. 2D)
   2. Fascicle sheath persistent
      3. Fascicles with 2 needles (count 15–20 fascicles, esp. on leading shoot)
         4. Needles 4–7 cm long, persisting 5–8 years; seed cones opening gradually, (3–)4–5.5 cm long · · · **3.** *P. contorta* var. *murrayana*
         4. Needles (7–)10–14(–16) cm long, persisting 2–3 years; seed cones serotinous, 5–7(–8) cm long
            . . . . **29.** *P. muricata* var. *muricata*
      3. Fascicles with 2–3 needles
         5. Seed cones usually in whorls of 2–5, serotinous, remaining closed for many years, (5–)8–15 cm long

6. Vegetative buds resinous; fascicles with 3 needles, rarely 2; seed cones attenuate · · · · · · · · · · · · · · · · · · **31.** *P. attenuata*
6. Vegetative buds not resinous; fascicles with 2 needles, sometimes 3 on leading shoots; seed cones ovoid or obliquely ovoid · · · · · · · · · · · · · · **30.** *P. radiata* var. *binata*

5. Seed cones solitary or in pairs, rarely in whorls of more than two, opening gradually or soon after maturity, 12–35 cm long
7. Seed cones very large and heavy, 20–35 cm long; apophyses and umbos of seed scales very strongly developed, very resinous · **33.** *P. coulteri*
7. Seed cones smaller, lighter, 12–17 cm long; apophyses and umbos of seed scales slightly raised, not resinous · **11.** *P. jeffreyi*

2. Fascicle sheath deciduous
8. Fascicles with 1 needle, rarely two · · · · · · · · · · **46.** *P. monophylla*
8. Fascicles with 3–5 needles, usually 4 · · · · · · · · · **47.** *P. quadrifolia*

Key to Species in NW Mexico (II)

(Sonora, Chihuahua, Sinaloa, Durango, Zacatecas)

1. Bases of cataphylls (pulvini) not decurrent (Fig. 1B); needles with a single vascular bundle and external resin ducts (microscope!); sheath of fascicle deciduous
2. Fascicles with 2–3 needles, rarely 4 or 5; seed cones small (3–5 cm), not longer than wide when open; seeds wingless when detached from the scale · **43.** *P. cembroides*
2. Fascicles with 5 needles; seed cones 10 cm long or longer; seeds with a short (sometimes vestigial) wing
3. Vegetative buds resinous; fascicle sheaths less than 15 mm long; seed cones 10–15 cm long · · · · · · · · · · · · **36.** *P. flexils* var. *reflexa*
3. Vegetative buds not resinous; fascicle sheaths 20–25 mm long; seed cones 12–30 cm long (or longer) · · · · · · · · · · · **37.** *P. strobiformis*

1. Bases of cataphylls (pulvini) decurrent (Fig. 1A); needles with two vascular bundles and variously positioned resin ducts (microscope!); sheath of fascicle persistent or deciduous
4. Sheath of fascicle deciduous · · · · · · · · · · · · · · · · · · · **1.** *P. leiophylla*
4. Sheath of fascicle persistent
5. Seed cones persistent, semi-serotinous, ovoid to globose when closed; resin ducts in the needles mostly septal (microscope!) · **20**. *P. oocarpa*
5. Seed cones falling, at least after a few years, opening when mature, obliquely ovoid to oblong when closed; resin ducts in the needles medial (microscope!)
6. Seed cones obliquely ovoid-oblong to attenuate, often curved, 15–35 cm long; fascicles with 5 needles, rarely 4 or 6 (count 14–20 fascicles) · **16.** *P. devoniana*
6. Seed cones ovoid or obliquely ovoid, smaller than 15 cm; fascicles with 3–5 needles, rarely 2
7. Peduncle deciduous with the seed cone, which falls intact · **28.** *P. teocote*

7. Peduncle persistent; the cone leaves some of the basal scales on the branch when it falls
    8. Seed cones 5–10(–12) cm long; apophyses of seed scales moderately raised (less than half as high as wide); needles rarely longer than 20 cm · · · · · · · · · · · · **10.** P. *arizonica*
    8. Seed cones 8–15 cm long; apophyses strongly raised (more than half as high as wide); needles (18–)20–35 cm long · **12.** *P. engelmannii*

Key to Species in W Mexico (III)

(Baja California Sur, Nayarit, S Zacatecas, Aguascalientes, Jalisco, Colima, Michoacán)

1. Fascicle sheaths deciduous
    2. Seed cones with ≤60, very widely spreading, very flexible scales; cones not longer than wide · · · · · · · · · · · · · · · · · · **43.** *P. cembroides*
    2. Seed cones with ≥60, inflexible or at least rigid scales; cones longer than wide
        3. Needles epistomatic, rarely a few stomata on the abaxial face; one vascular bundle in the needle
            4. Fascicles sheaths with distinct, individually falling scales; seed cones cylindrical, 12–30(–60) cm long; apophyses of seed scales, at least the proximal ones, recurved or reflexed · **37.** *P. strobiformis*
            4. Fascicles sheaths with basally connate scales recoiling to form a rosette before they fall; seed cones not cylindrical; apophyses not recurved or reflexed
                5. Fascicles with (3–)4–5 needles; seed cones 10–15 cm long, with thin, lignified but bendable scales; seeds small (ca. 8 mm long), winged · · · · · · · · · · · · · · · · **39.** *P. rzedowskii*
                5. Fascicles with 5 needles, rarely 3–4; seed cones (15–)17–25(–27) cm long, with thick, inflexible scales; seeds 20–28 mm long, wingless · · · · · · · **40.** *P. maximartinezii*
        3. Needles amphistomatic; two vascular bundles in the needle
            6. Needles (4–)6–15(–17) cm long, spreading; seed scales with a conspicuous narrow band visible around the umbo · **1.** *P. leiophylla*
            6. Needles (15–)20–30(–40+) cm long, pendulous; seed scales lacking a band around the umbo · · · · · · · · **19.** *P. lumholtzii*
1. Fascicle sheaths persistent
    7. Fascicles with 3 needles, sometimes a few 2–5 (count 15–20 fascicles); seed cones (2–)3–6(–7) cm long
        8. Seed cones ovoid to subglobose when closed, shorter than wide when opened · · · · · · · · · · · · · · · · **20b.** *P. oocarpa* var. *trifoliata*
        8. Seed cones ovoid or obliquely ovoid when closed, longer than wide when opened
            9. Needles slender, lax, (10–)15–20 cm long, 0.7–0.9 mm wide; seed cones (2–)3–3.5(–4) cm long · · · · · · · · · · **4.** *P. herrerae*
            9. Needles rigid, (7–)10–15(–18) cm long, 1–1.4 mm wide; seed cones (3–)4–6(–7) cm long · · · · · · · · · · · · · **28.** *P. teocote*

7. Fascicles with (4–)5 needles, rarely 3 or 6; seed cones (4–)5–10(–12) cm long or much larger
  10. Seed cones (4–)5–10(–12) cm long; seed scales with flat or slightly raised apophyses (Fig. 2E, F)
    11. Needles (8–)10–25 cm long
      12. Seed cones ovoid to subglobose when closed, shorter than wide when opened
        13. Cones falling with the proximal seed scales missing; needles (8–)10–16 cm long, 0.5–0.8 mm wide; resin ducts in the needles internal (microscope!) · **21.** *P. praetermissa*
        13. Cones persistent, intact when falling; needles (11–)14–25 cm long; resin ducts septal (microscope!) · **20a.** *P. oocarpa* var. *oocarpa*
      12. Seed cones ovoid-oblong to attenuate or obliquely ovoid when closed, longer than wide when opened
        14. Proximal seed scales not parting when the cone opens; resin ducts in the needles septal (microscope!) · **23.** *P. jaliscana*
        14. Proximal seed scales parting when the cone opens; resin ducts in the needles medial (microscope!)
          15. Seed cones with ca. 150–200 scales; apophyses of seed scales more or less flat, weakly keeled, often purplish-black · · · · · · · · · · · · · · · **13.** *P. hartwegii*
          15. Seed cones with ca. 90–120 scales; apophyses of seed scales (slightly) raised, prominently keeled, ochraceous to light (reddish-)brown · **25.** *P. durangensis*
    11. Needles 20–35 cm long
      16. Needles 0.6–1(–1.1) mm wide, drooping; seed scales thin woody, spreading 900 or reflexed when the cone is open; hypodermis with numerous intrusions into the mesophyll, some connecting with the endodermis (microscope!) · **18.** *P. maximinoi*
      16. Needles 0.7–1.2 mm wide, spreading or drooping; seed scales woody, spreading less than 900 when the cone is open; hypodermis without or with few intrusions into the mesophyll (microscope!) · · · · · · · · · · **17.** *P. douglasiana*
  10. Seed cones (7–)8–35 cm long; at least several seed scales with prominently raised apophyses (rarely all nearly flat) (Fig. 2G, H)
    17. Fascicle sheaths 30–40 mm long, resinous; needles 1.1–1.6 mm wide; outer walls of endodermal cells thin; seed cones 15–35 cm long · **16.** *P. devoniana*
    17. Fascicle sheaths 20–30(–35) mm long, usually not resinous; needles 0.8–1.3 mm wide; outer walls of endodermal cells thickened; seed cones 8–20 cm long
      18. Needles with connate vascular bundles; seed cones usually obliquely ovoid when closed · · · · · · **14.** *P. pseudostrobus*

18. Needles with separate vascular bundles; seed cones usually ovoid-oblong to attenuate when closed · **15.** *P. montezumae*

Key to Species in NE & E Mexico (IV)

(E Chihuahua, Coahuila, Nuevo León, Tamaulipas, N Zacatecas, San Luis Potosí, Guanajuato, Querétaro, Hidalgo, Tlaxcala, N Veracruz, N Puebla)

1. Fascicle sheaths deciduous
   2. Seed cones with ≤60, very widely spreading, very flexible scales; cones not longer than wide
      3. Fascicles with 5 needles (very rarely 4 or 6, count 15–20 fascicles); a low, spreading shrub · · · · · · · · · · · · · · · · · **44.** *P. culminicola*
      3. Fascicles with 2–4(–5) needles; a (small) tree
         4. Scales of fascicle sheaths recoiling strongly before falling; needles in fascicles of (2–)3(–4), rarely five, less than 1 mm wide · **43.** *P. cembroides*
         4. Scales of fascicle sheaths not recoiling; fascicles with 2(–3) needles, 0.8–1.1 mm wide · · · · · · · · · · · · · · · · **45.** *P. remota*
   2. Seed cones with ≥60 scales, spreading ≤90°, rigid, or if flexible, spreading very little; cones longer than wide
      5. Fascicles with 3 needles, rarely 4 (count 15–20 fascicles); seed cones oblong, irregular, up to 10–12 cm long; seed scales very flexible; seeds wingless when detached from the scale · **42.** *P. pinceana*
      5. Fascicles with (2–)3–5(–6) needles; seed cones ovoid or cylindrical, regular; seed scales rigid; seeds winged
         6. Fascicles with 5 needles, rarely 6; seed cones cylindrical
            7. Seed wing shorter than seed or vestigial (Fig. 2Mb); needles 6–11 cm long, 0.8–1.2 mm wide
               8. Vegetative buds resinous; fascicle sheaths less than 15 mm long; seed cones 10–15 cm long · **36.** *P. flexilis* var. *reflexa*
               8. Vegetative buds not resinous; fascicle sheaths 20–25 mm long; seed cones 12–30 cm long (or longer) · **37.** *P. strobiformis*
            7. Seed wings longer than seed (Fig. 2I, J); needles (6–)8–15 cm long, 0.6–1 mm wide · · · · · · · · · · · **34.** *P. ayacahuite*
         6. Fascicles with (2–)3–5(–6) needles; seed cones ovoid · **1.** *P. leiophylla*
1. Fascicle sheaths persistent
   9. Bases of cataphylls (pulvini) not decurrent (Fig. 1B); fascicles with 3 needles, but connate, appearing as one; seed cones on thick, long and curved, persistent peduncles; seeds wingless when detached from the scale · **41.** *P. nelsonii*
   9. Bases of cataphylls (pulvini) decurrent (Fig. 1A); fascicles with (2–)3–5(–6) needles, separate; seed cones on (relatively) short peduncles; seeds winged

10. Needles (7–)9–15(–18) cm long; seed cones (semi-)serotinous; apophyses of seed scales flat or only slightly raised (Fig. 2E, F)
    11. Seed cones conspicuously pedunculate, in whorls of 1–3, (3–)4–6(–7) cm long; fascicles with 3 needles, but occasionally 2–5 (count 15–20 fascicles) ··············· **28.** *P. teocote*
    11. Seed cones appearing sessile, in whorls of 3–8, (6–)8–13(–15) cm long; fascicles with 3 needles ··········· **32.** *P. greggii*
10. Needles usually longer than 15 cm; seed cones opening soon after maturity; apophyses of seed scales raised, or at least prominently keeled (Fig. 2F, G)
    12. Needles very slender, drooping in two rows, 0.7–0.9(–1) mm wide, in fascicles of 3–4(–5) ··············· **22.** *P. patula*
    12. Needles wider than 1 mm; if 0.8–1 mm, not drooping in two rows; predominantly in fascicles of five
        13. Fascicles with (2–)3–4(–5) needles (count 15–20 fascicles).
            14. Umbo of seed scales with a persistent spine or prickle; seed cones 8–15 cm long; needles 20–35 cm long ························· **12.** *P. engelmannii*
            14. Umbo of seed scales with a tiny, deciduous prickle (or prickle absent); seed cones (4.5–)5–10(–14) cm long; needles 14–25 cm long ················· **10c.** *P. arizonica* var. *stormiae*
        13. Fascicles with (3–)4–5(–6) needles, predominantly 5
            15. Needles (6–)10–17(–22) cm long; seed scales of cones relatively thin woody, slightly flexible ··························· **13.** *P. hartwegii*
            15. Needles (15–)20–40(–45) cm long; seed scales of cones relatively thick woody, inflexible
                16. Fascicle sheaths 30–40 mm long, resinous; needles 1.1–1.6 mm wide; outer walls of endodermal cells thin (microscope!); seed cones 15–35 cm long ························ **16.** *P. devoniana*
                16. Fascicle sheaths 20–30(–35) mm long, usually not resinous; needles 0.8–1.3 mm wide; outer walls of endodermal cells thickened (microscope!); seed cones 8–20 cm long
                    17. Needles with connate vascular bundles; seed cones usually obliquely ovoid when closed ···················· **14.** *P. pseudostrobus*
                    17. Needles with separate vascular bundles; seed cones usually ovoid-oblong to attenuate when closed ················ **15.** *P. montezumae*

Key to Species of Central & S Mexico (V)

(E Michoacán, México, D.F., Morelos, Hidalgo, Tlaxcala, Puebla, Guerrero, Oaxaca, S Veracruz, Chiapas)

1. Fascicle sheaths deciduous
    2. Seed cones with ≤60, very widely spreading, very flexible scales; cones not longer than wide ················· **43.** *P. cembroides*

2. Seed cones with ≥60, rigid scales, spreading ≤90°; cones longer than wide
   3. Fascicles with (2–)3–5(–6) needles (count 15–20 fascicles); seed cones ovoid, (4–)5–7(–8) cm long; umbo on seed scales dorsal (Fig. 2Da, c) · · · · · · · · · · · · · · · · **1a.** *P. leiophylla* var. *leiophylla*
   3. Fascicles with 5 needles; seed cones cylindrical, (6–)8–40(–50) cm long; umbo on seed scales terminal. (Fig. 2Ca)
      4. Apophyses, at least those of proximal seed scales, recurved or reflexed; seed cones usually 15–40 cm long · **34.** *P. ayacahuite*
      4. Apophyses of seed scales not reflexed; seed cones usually 8–16 cm long, rarely longer · · · **38.** *P. strobus* var. *chiapensis*

1. Fascicle sheaths persistent
   5. Seed cones asymmetrical, oblique at base or curved, 4.5–35 cm long; seed scales opening to 90° or more; apophysis nearly flat to prominently raised (Fig. 2E–G)
      6. Needles relatively short, (6–)10–17(–22) cm long, spreading; apophyses of seed scales nearly flat or slightly raised (Fig 2E, F), often purplish-black · **13.** *P. hartwegii*
      6. Needles very long, 20–45 cm, spreading or drooping; apophyses of seed scales nearly flat or prominently raised (Fig. 2E–G)
         7. Cones 5–10(–12) cm long; falling intact (peduncle deciduous with the cone)
            8. Seed scales of cones thin woody, flexible, in open cones usually strongly recurved; needles very slender, drooping, 0.6–1 mm wide · · · · · · · · · · · · · · · · · · **18.** *P. maximinoi*
            8. Seed scales of cones not flexible, not strongly recurved in open cones; needles drooping or spreading, 0.7–1.2 mm wide · **17.** *P. douglasiana*
         7. Cones (7–)10–35 cm long; leaving a few proximal scales on the branch when falling
            9. Fascicle sheaths 30–40 mm long, resinous; needles 1.1–1.6 mm wide; outer walls of endodermal cells thin; seed cones 15–35 cm long · · · · · · · · · · · · · · **16.** *P. devoniana*
            9. Fascicle sheaths 20–30(–35) mm long, usually not resinous; needles 0.8–1.3 mm wide; outer walls of endodermal cells thickened; seed cones 8–20 cm long
               10. Needles with connate vascular bundles (microscope!); seed cones usually obliquely ovoid when closed · **14.** *P. pseudostrobus*
               10. Needles with separate vascular bundles (microscope!); seed cones usually ovoid-oblong to attenuate when closed · **15.** *P. montezumae*
   5. Seed cones symmetrical, ovoid or ovoid-oblong, sometimes slightly oblique, (4–)5–10(–12) cm long; seed scales usually spreading less than 90°; apophysis flat or slightly raised (Fig. 2E, F)
      11. Cones broadly ovoid to subglobose when closed
         12. Cones semi-serotinous (only the distal seed scales parting), remaining on the tree; needles 0.8–1.6 mm wide; resin ducts in the needles septal (microscope!) · · · · · · · · **20.** *P. oocarpa*

12. Cones opening completely when mature, falling after 1–3 years; needles 0.7–1 mm wide; resin ducts in the needles medial (microscope!) · · · · · · · · · · · · · · · **24.** *P. tecunumanii*

11. Cones ovoid to attenuate when closed
    13. Needles with 10 or more lines of stomata on tha abaxial face; resin ducts in the needles internal (microscope!)
        14. Cones semi-serotinous, persistent, leaving some basal scales on the branch when falling; umbo of seed scales flat or depressed · **27.** *P. pringlei*
        14. Cones opening quickly at maturity, falling soon with peduncles attached; umbo of seed scales prominently raised · **26.** *P. lawsonii*
    13. Needles with 3–7(–10) lines of stomata on the abaxial face; resin ducts in the needles medial (microscope!)
        15. Needles (7–)10–15(–18) cm long, 1–1.4 mm wide · **28.** *P. teocote*
        15. Needles (11–)15–25(–30) cm long, 0.7–0.9(–1) mm wide · **22.** *P. patula*

Key to Species of Mesoamerica (VI)

(Chiapas, Quintana Roo, Belize, Guatemala, El Salvador, Honduras, Nicaragua)

1. Fascicle sheaths deciduous
    2. Apophyses, at least those of proximal seed scales, recurved or reflexed; seed cones usually 15–40 cm long · · · · · **34.** *P. ayacahuite*
    2. Apophyses of seed scales not reflexed; seed cones usually 8–16 cm long, rarely longer · · · · · · · · · · · · · · · **38.** *P. strobus* var. *chiapensis*
1. Fascicle sheaths persistent
    3. Long shoots usually multinodal; vegetative buds usually resinous; needles in fascicles of three (rarely two or four); resin ducts internal (microscope!) · · · · · · · **5c.** *P. caribaea* var. *hondurensis*
    3. Long shoots uninodal; vegetative buds usually not resinous; needles in fascicles of 4–5, rarely three or six; resin ducts variously positioned (microscope!)
        4. Seed cones asymmetrical, oblique at base or curved, 4.5–35 cm long; seed scales opening to 90° or more; apophysis nearly flat to prominently raised (Fig. 2E–G)
            5. Needles relatively short, (6–)10–17(–22) cm long, spreading; apophyses of seed scales nearly flat or slightly raised (Fig. 2E, F), often purplish-black · · **13.** *P. hartwegii*
            5. Needles very long, 20–45 cm, spreading or drooping; apophyses of seed scales nearly flat or prominently raised. (Fig. 2E, G)
                6. Cones 5–10(–12) cm long; falling intact (peduncle deciduous with the cone); seed scales thin woody, flexible, in open cones usually strongly recurved; needles very slender, drooping, 0.6–1 mm wide · **18.** *P. maximinoi*

6. Cones (7–)10–35 cm long; leaving a few proximal scales on the branch when falling; seed scales thick woody, rigid, usually not recurved; needles 0.8–1.6 mm wide
   7. Fascicle sheaths 30–40 mm long, resinous; needles 1.1–1.6 mm wide; outer walls of endodermal cells thin; seed cones 15–35 cm long
      · **16.** *P. devoniana*
   7. Fascicle sheaths 20–30(–35) mm long, usually not resinous; needles 0.8–1.3 mm wide; outer walls of endodermal cells thickened; seed cones 8–20 cm long
      8. Needles with connate vascular bundles; seed cones usually obliquely ovoid when closed
         · **14.** *P. pseudostrobus*
      8. Needles with separate vascular bundles; seed cones usually ovoid-oblong to attenuate when closed · · · · · · · · · · · · · · · · **15.** *P. montezumae*

4. Seed cones symmetrical, ovoid or ovoid-oblong, sometimes slightly oblique, (4–)5–10(–12) cm long; seed scales usually spreading less than 90°; apophysis flat or slightly raised.(Fig. 2E, F)
   9. Cones semi-serotinous (only the distal seed scales parting), remaining on the tree; needles 0.8–1.6 mm wide; resin ducts in the needles septal (microscope!)
      · **20.** *P. oocarpa*
   9. Cones opening completely when mature, falling after 1–3 years; needles 0.7–1 mm wide; resin ducts in the needles medial (microscope!) · · · · · **4.** *P. tecunumanii*

KEYS BY MORPHOLOGICAL GROUPS

The keys by morphological groups have a limited application because together they do not cover all the species as is the case with the regional keys. This is because there are species of pines which lack any of the characters used here to define the morphological groups. If a pine in the field cannot be determined by any of the keys that follow, one should work with the appropriate regional key instead. Species not included in the following keys are: *P. arizonica*, *P. lawsonii*, *P. leiophylla*, *P. maximartinezii*, *P. monophylla*, *P. ponderosa* var. *scopulorum*, *P. pringlei* and *P. rzedowskii*.

Key to the morphological groups

1. Fascicles with 2 needles (rarely a few fascicles with 3) · · · · · · · **group 1**
1. Fascicles with more than 2 needles (sometimes a few with 2)
   2. Fascicles with 3 needles (rarely a few with 2 or 4) · · · · · · · · **group 2**
   2. Fascicles with (2–) 3, 4 and/or 5 (sometimes more) needles

3. Fascicles with a variable number of needles (count many fascicles to determine this!), or if only with 5, then fascicle sheath recoiling before it falls from the fascicle
   4. Cones small, with ≤60, widely spreading scales and wingless seeds · **group 3**
   4. Cones small or large, with ≥60 more or less spreading scales and winged seeds
      5. Cones shorter than ca. 6 cm · · · · · · · · · · · · · · · · · · · **group 4**
      5. Cones ca. 6–15 cm long · **group 5**
3. Fascicles with 5 needles (rarely 4 or 6), with a deciduous or persistent sheath of which the scales do not recoil
   6. Fascicle sheaths deciduous, needles not longer than 18 cm; cone scales with terminal umbos · **group 6**
   6. Fascicle sheaths persistent, needles longer than 18 cm; cone scales with dorsal umbos · **group 7**

Key to Species in group 1

1. Needles 4–7 cm long, 1.2–2.0 mm wide, margins entire or weakly serrulate; cones small, (3–)4–5.5 × 3–4 cm when open · **3.** *P. contorta* var. *murrayana*
1. Needles longer than 7 cm, margins serrulate; cones longer than 5 cm
   2. Cones variable, with slightly raised to extremely elongated apophyses on one side of the cone, umbo armed with a sharp prickle · **29.** *P. muricata*
   2. Cones more uniform, with slightly raised or thickened apophyses on one side of the cone, with a flat umbo · · · · **30.** *P. radiata* var. *binata*

Key to Species in group 2

1. Cones very persistent, sessile, usually in whorls and remaining closed for many years
   2. Cone scales with conical, curved apophyses on one side near the base of the cone · **31.** *P. attenuata*
   2. Cone scales with flat or slightly raised apophyses on all sides of the cone · **32.** *P. greggii*
1. Cones (eventually) deciduous, usually pedunculate, opening soon or gradually
   3. Needles (15–)20–30(–40+) cm long, very pendulous, fascicle sheaths deciduous, the long brown scales curling before they fall · **19.** *P. lumholtzii*
   3. Needles shorter, or if long not pendulous, fascicle sheaths persistent, or if deciduous with short, inconspicuous scales
      4. Foliage branches all pendulous, fascicle sheaths deciduous; shrubs or small, bushy trees · **42.** *P. pinceana*
      4. Foliage branches mostly spreading, fascicle sheaths persistent
         5. Needles in fascicles all free; bark not whitish
            6. Needles thick (1.5–2.2 mm wide), very rigid, spreading

7. Cones numerous on all trees, (10–)12–17 × 9–14 cm when open, with thin scales and slightly raised apophyses, falling soon · **11.** *P. jeffreyi*
7. Only a few cones (if any) per tree, massive, 20–35 × 15–20 cm when open, with thick scales and elongated, hooked apophyses and umbos, remaining some years on the tree · **33.** *P. coulteri*

6. Needles less than 1.6 mm wide, lax or rigid, drooping or spreading
8. Needles slender, 0.7–0.9 mm wide, lax, drooping; cones very small, (2–)3–3.5(–4) × 2–3.5 cm when open, falling soon · **4.** *P. herrerae*
8. Needles 1.2–1.6 mm wide, rigid, spreading, cones usually larger, remaining some years on the tree · **20b.** *P. oocarpa* var. *trifoliata*

5. Needles in fascicles connate, appearing as one; shrub or small tree with whitish bark · **41.** *P. nelsonii*

Key to Species in group 3

1. Fascicles with 5 needles (very rarely 4 or 6); a low, spreading shrub · **44.** *P. culminicola*
1. Fascicles with 2–4(–5) needles; a small tree
2. Scales of fascicle sheaths recoiling strongly before falling; needles less than 1 mm wide · **43.** *P. cembroides*
2. Scales of fascicle sheaths not recoiling; needles 0.8–1.5 mm wide
3. Needles in fascicles of 2(–3), 0.8–1.1 mm wide · · · · · · · **45.** *P. remota*
3. Needles in fascicles of (3–)4(–5), 1–1.5 mm wide · · **47.** *P. quadrifolia*

Key to Species in group 4

1. Fascicles with 3, rarely 2, 4 or 5 needles (count many fascicles)
2. Fascicle sheaths long, deciduous, needles very pendulous, (15–) 20–30(–40) cm long · **19.** *P. lumholtzii*
2. Fascicle sheaths short, persistent, needles spreading or drooping, not pendulous, (7–)10–15(–18) cm long · · · · · · · · · · · · · · · · **28.** *P. teocote.*
1. Fascicles with (3–)4–5 needles
3. Needles (8–)10–16 cm long, 0.5–0.8 mm wide, very slender, lax; cones broadly ovoid and smooth when closed, with a slender peduncle, often loosing basal scales when they fall · · · · · · · **21.** *P. praetermissa*
3. Needles (14–)16–18(–25) cm long, 0.7–1.0(–1.1) mm wide, straight, lax; cones ovoid to broadly ovoid when closed, with a stout, curved peduncle, not loosing basal scales · · · · · · · · · · · · **24.** *P. tecunumanii*

Key to Species in group 5

1. Fascicles with (4–)5–6(–7, rarely 8) needles; cones 5–9(–11) × 4–6(–7) cm when open, cone scales thick woody, rigid ······· **25.** *P. durangensis*
1. Fascicles with (2–)3–4 or (3–)4–5 (rarely 6) needles (count 20–25 fascicles)
   2. Fascicles predominantly with 3 needles, sometimes more, rarely 2
      3. Cones leaving a few basal scales when falling; cone scales with a persistent, sharp prickle on the umbo; needles 1.5–2.0 mm wide, usually very rigid ······················ **12.** *P. engelmannii*
      3. Cones falling with the peduncle; cone scales without a persistent prickle on the umbo; needles (1.2–)1.4–1.8 mm wide, rigid ························· **5.** *P. caribaea* var. *hondurensis*
   2. Fascicles with 3–5 needles, rarely 6
      4. Cones to 10(–12) cm long, remaining closed near the base and opening slowly
         5. Fascicles with 3–4(–5) needles; cones in whorls of 2 or more, nearly sessile, persistent ····················· **22.** *P. patula*
         5. Fascicles with (4–)5, rarely 3 needles; cones solitary or in whorls of 2–3 on curved peduncles ··············· **23.** *P. jaliscana*
      4. Cones 8–15(–20) cm long, opening soon and almost completely
         6. Needles (6–)10–17(–22) cm long, straight or curved, rigid; cone scales thin, with a more or less flat apophysis, often purplish-blackish ···························· **13.** *P. hartwegii*
         6. Needles (20–)25–35 cm long, straight, lax or more rigid; cone scales thin or thick, with a raised apophysis, brown ································ **15.** *P. montezumae*

Key to Species in group 6

1. Cones (6–)8–16(–25) cm long, scales (except sometimes near the base of the cone) not recurved
   2. Needle fascicles persisting (3–)5–6 years; cone scales with thick apophyses; seed with a rudimentary wing or wing absent on the seed ······························· **36.** *P. flexilis* var. *reflexa*
   2. Needle fascicles persisting 2–3 years; cone scales with thin apophyses; seed with a wing 20–30 × 6–9 mm adnate to the seed ····························· **38.** *P. strobus* var. *chiapensis*
1. Cones (10–)15–45(–60 cm long, scales recurved or straight, if straight cones usually longer than 25 cm
   3. Cones 25–45 × 8–14 cm when open, scales all straight with thick apophyses ···························· **35.** *P. lambertiana*
   3. Cones variable in size, even on a single tree, scales recurved, at least the tips and those near the base of the cone
      4. Seeds with short or rudimentary wings (see the seed wing marks on the inside of the seed scales if the seeds have fallen from the cone) ······························ **37.** *P. strobiformis*
      4. Seeds with well developed wings about 2× longer than the seeds ································ **34.** *P. ayacahuite*

Key to Species in group 7

1. Cones broadly ovoid to subglobose when closed, with a flattened base and 3–8(–10) × 3–9(–12) cm when open, persisting for several years . . . . **20.** *P. oocarpa*
1. Cones longer than wide when closed, usually asymmetrical, deciduous.
   2. Needles very thin and lax, 0.6–1.0(–1.1) mm wide, drooping to pendulous; cone scales thin, often reflexed in open cones . . . . **18.** *P. maximinoi*
   2. Needles usually thicker, drooping or spreading; cone scales thick woody, not reflexed in open cones
      3. Fascicle sheaths very long, up to 40 mm, resinous, needles 1.1–1.6 mm wide, lustrous green; cones 15–35 × 8–15 cm when open, often curved . . . . **16.** *P. devoniana*
      3. Fascicle sheaths up to 30 mm long, not resinous, needles 0.7–1.3 mm wide, dull green or glaucous-green; cones up to 16 cm long
         4. Cones 7–10 × 5–7 cm when open, (broad) ovoid, regular, scales with slightly raised apophysis; thick intrusions of hypodermal cells within the needles (microscope!) . . . . . **17.** *P. douglasiana*
         4. Cones highly variable, 7–16 × 6–13 cm when open, usually oblique, scales with slightly or extremely raised apophysis; no intrusions of hypodermal cells within the needles . . . . **14.** *P. pseudostrobus*

1 Pinus leiophylla

Pinus leiophylla Schlechtendal & Chamisso
var. *leiophylla*

Local names: Ocote, Pino chino, Pino prieto (also var. *chihuahuana*).

Habit, trunk: tree to 20–30 m tall, d.b.h. 50–85 cm, trunk straight.

Bark: very thick on the trunk, scaly, with deep fissures and irregular but elongated plates, dark grey-brown.

Foliage twigs: reddish-brown, sometimes glaucous, soon grey-brown; needle fascicles drooping or spreading, persisting 2–3 years.

Needles: in fascicles of (4–)5(–6), 4 more often than 6, (6–)8–15(–17) cm long, 0.5–0.9 mm wide, lax, sometimes more rigid.

Cones: solitary or in whorls of 2–5, maturing in three seasons (therefore often cones of three distinct ages on the tree), symmetrical, ovoid, (4–)5–7(–8) × (3–)4–5.5 cm when open.

Cone scales: 50–70, opening soon; apophysis with distinct section of second season's growth around the darker umbo.

Seeds: 3–4(–5) mm long, with articulate wing 10–18 × 4–8 mm, lighter coloured than the seed.

Habitat: montane to high montane pine and pine-oak forests, usually on deep, well drained soils; this species is a common constituent of both forest types.

Distribution: MEXICO: NE Sonora, W Chihuahua, Durango, Nayarit, Zacatecas, Jalisco, Michoacán, Mexico, D.F., Hidalgo, Morelos, Tlaxcala, Puebla, Veracruz, Guerrero and Oaxaca.

Altitude: (1500–)1900–2900(–3300) m.

Notes: *Pinus leiophylla* is the only species in the region in which the cones take three seasons ("years") to grow to maturity. It is one of the few species of pine that regrows from stumps. Occasionally a form of *P. leiophylla* is found with very lax leaves as in *P. lumholzii*.

Related species, subspecies or varieties: In the north *P. leiophylla* var. *leiophylla* is gradually replaced by *P. leiophylla* var. *chihuahuana* (Engelmann) Shaw (syn. *P. chihuahuana*), although var. *leiophylla* reaches NE Sonora. The variety *chihuahuana* may be a smaller tree due to soil and climate conditions and differs only in the shorter, thicker, rigid leaves ((4–)6–12(–14) cm x 0.9–1.3(–1.5) mm) in fascicles of (2–)3(–4), rarely 5.

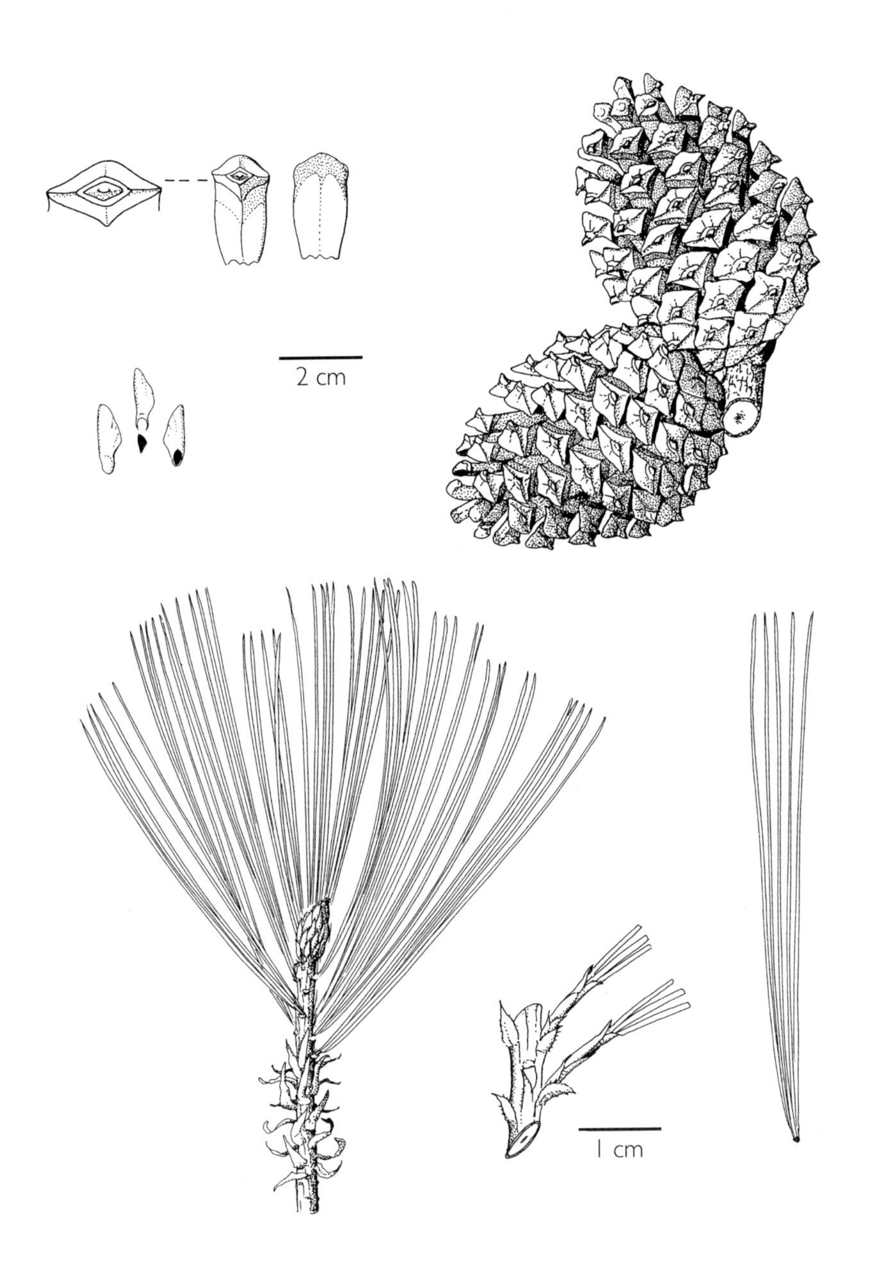
2 cm
1 cm

3 Pinus contorta var. murrayana

Pinus contorta J. C. Loudon var. *murrayana* (Balfour) Engelmann

Local names: Pino

Habit, trunk: tree to 25–33 m tall, d.b.h. 100–150(200) cm, trunk straight, sometimes forked.

Bark: thin, scaly, with numerous small, flaking plates, light orange- or pink-brown, later greyish.

Foliage twigs: young shoots glaucous, rough; leading shoots often multinodal; needle fascicles erect or spreading, persisting 5–8 years.

Needles: in fascicles of 2, 4–7 cm long, 1.2–2.0 mm wide, entire or sparsely serrulate, usually curved and contorted, rigid.

Cones: solitary or in whorls of 2–5 on short peduncles, ovoid with an oblique base, (3–)4–5.5 × 3–4 cm when open.

Cone scales: 90–110, opening gradually; apophysis transversely keeled, often thickened on one side near the base of the cone; umbo with a prominent, persistent prickle.

Seeds: 4–5 mm long, greyish-brown, with articulate wing 8–12 × 4–5 mm, yellowish-brown.

Habitat: high montane pine and mixed conifer forests, on deep, well drained soils as well as rocky slopes.

Distribution: MEXICO: only on the higher parts of the Sierra San Pedro Martír in Baja California Norte; it is the southernmost occurrence of this variety, which extends through the Pacific states of the U.S.A.

Altitude: 2300–3000+ m.

Notes: Most of the Mexican trees of this variety of *Pinus contorta* grow inside the Parque Nacional San Pedro Martír and are therefore protected from exploitation.

Related species, subspecies or varieties: No other varieties of *Pinus contorta* occur in Mexico; Martínez (1948) erroneously identified the trees of Baja California as *P. contorta* var. *latifolia*, which occurs in the Rocky Mountains.

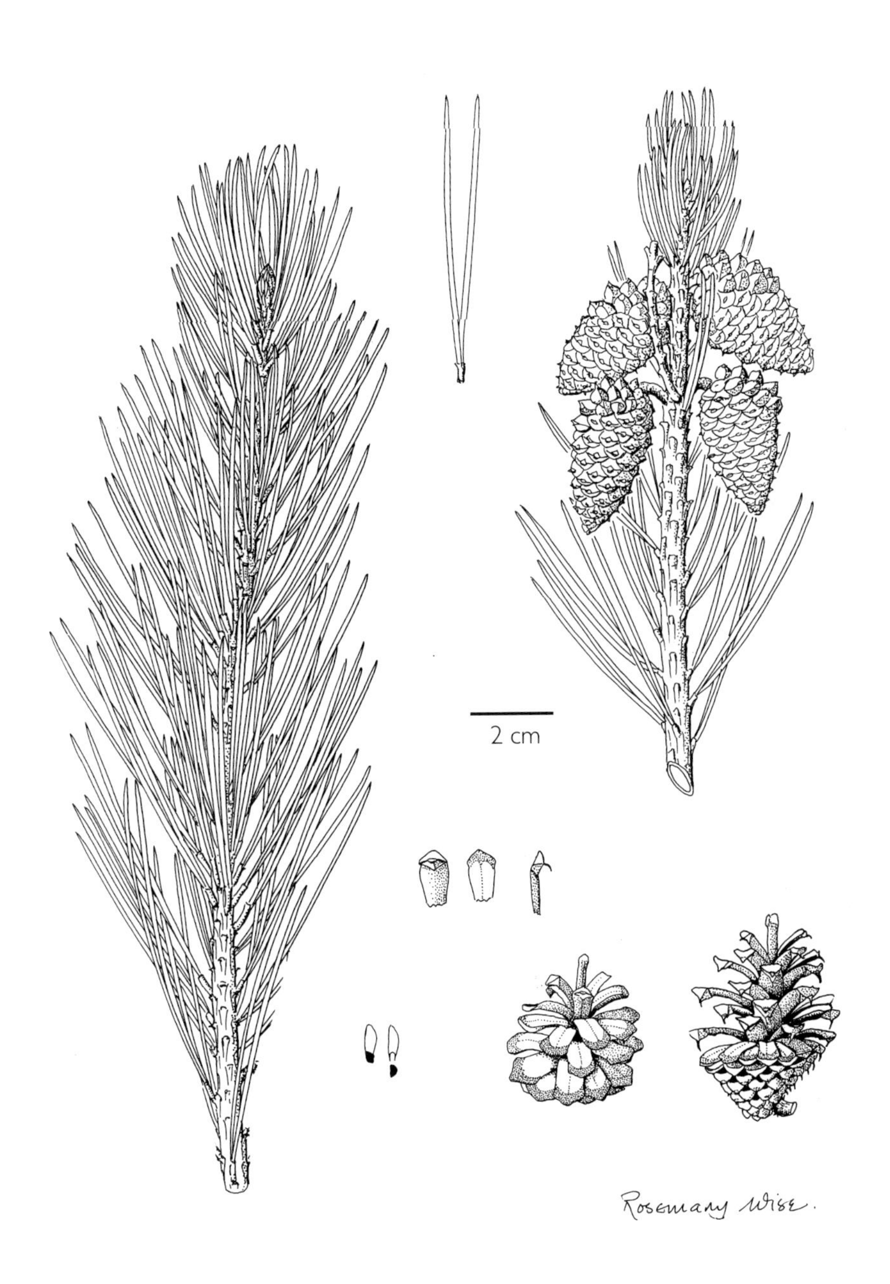
2 cm
Rosemary Wise.

❹ Pinus herrerae

Pinus herrerae Martínez

Local names: Ocote, Pino chino

Habit, trunk: tree to 30–35 m tall, d.b.h. 75–100 cm, trunk straight, sometimes tortuous.

Bark: thick on the trunk, with scaly plates and shallow, longitudinal fissures, reddish-brown to grey-brown.

Foliage twigs: orange-brown, with long decurrent leaf bases (fascicle bases); needle fascicles drooping or spreading, persisting 3 years.

Needles: in fascicles of 3, (10–)15–20 cm long, 0.7–0.9 mm wide, slender, lax.

Cones: solitary or in pairs, rarely in whorls of 3, on distinct peduncles, ovoid, very small, (2–)3–3.5(–4) × 2–3.5 cm when open.

Cone scales: 50–80, opening soon; apophysis slightly raised, with a small, mucronate umbo.

Seeds: 2.5–4 × 2–3 mm, with articulate wing 5–8 × 3–5 mm.

Habitat: the mesic montane forest belt in mixed pine and pine-oak forest; locally this pine grows with *Pseudotsuga* (Douglas fir).

Distribution: MEXICO: SW Chihuahua, Sinaloa, Durango, W and S Jalisco, Michoacán and Guerrero.

Altitude: (1100–)1500–2600 m.

Notes: This pine has the smallest cones of any Mexican pine.

Similar species: This species is in many characters similar to *P. teocote* (No. 28), yet distinct in a few constant characters. The needles are more slender, lax and usually longer and the cones are smaller. Anatomical details in the leaves (see Farjon & Styles, Flora Neotropica Monograph 75) further distinguish these two pines, which may not even be closely related.

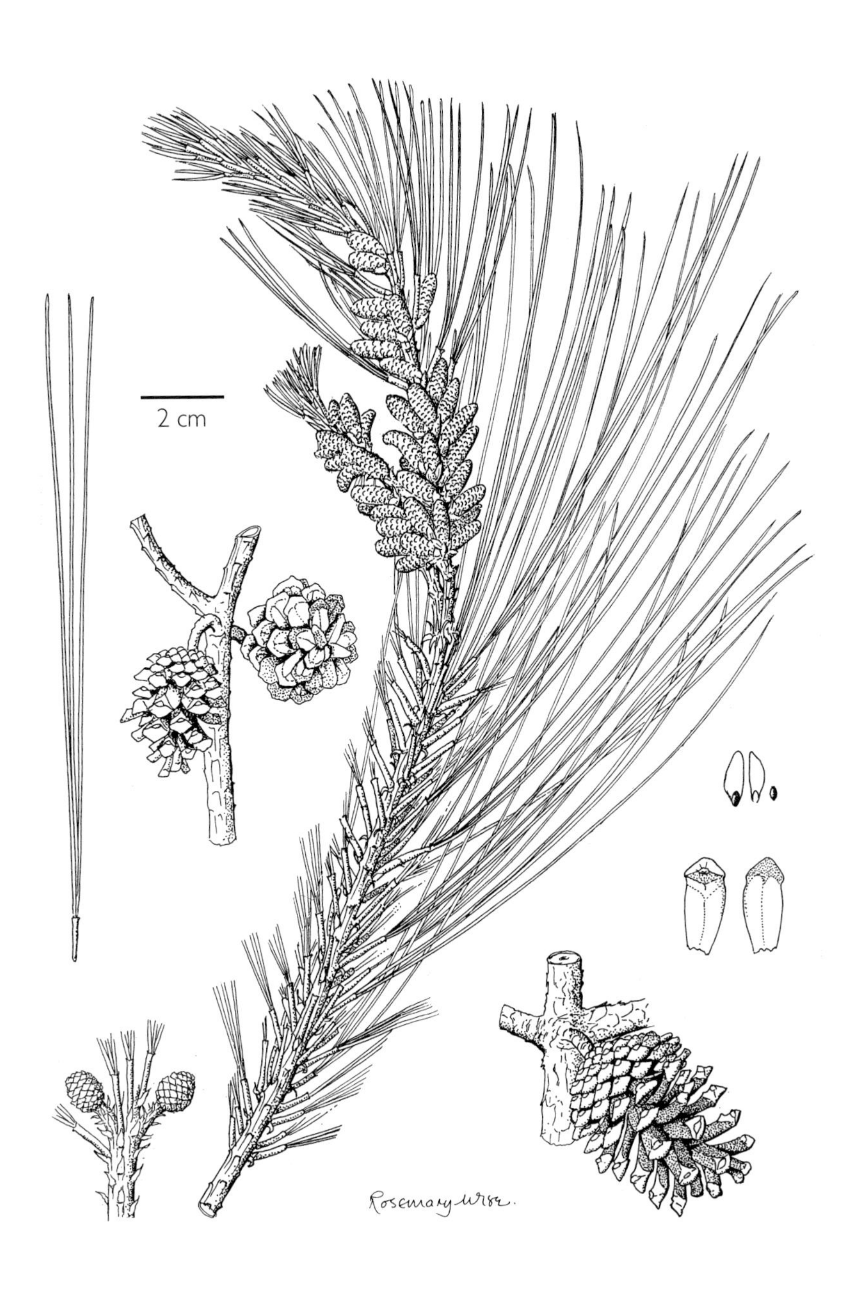
2 cm
Rosemary Wise 82.

5 Pinus caribaea var. hondurensis

Pinus caribaea Morelet var. *hondurensis* (Sénéclauze) W. H. Barrett & Golfari

Synonyms: *P. hondurensis* Loock

Local names: Ocote, Pino

Habit, trunk: tree to 20–35(–40) m tall, d.b.h. 60–100 cm, trunk straight.

Bark: rough, scaly, on lower part of trunk breaking into irregularly square plates divided by shallow or deep fissures, inner bark reddish-brown, outer bark grey-brown.

Foliage twigs: (main) shoots multinodal, rough, resinous; needle fascicles spreading, persisting 3 years.

Needles: in fascicles of 3 (rarely 2, 4, very rarely 5), (12–)16–28 cm long, (1.2–)1.4–1.8 mm wide, straight, rigid.

Cones: often in several whorls in one season's growth of a branch, in pairs or whorls of 3–5(–8) on curved peduncles, deciduous, (4–)5–12(–13) × 3.5–7 cm when open.

Cone scales: 120–170, opening soon; apophysis (slightly) raised, less so near the cone base, irregularly rhombic in outline, chestnut brown, lustrous, with a raised umbo.

Seeds: 5–7 × 2.5–3.5 mm, with an articulate or adnate wing often enclosing the seed on one side and 10–20 × 5–8 mm, wing usually lighter coloured than the seed.

Habitat: mainly lowland coastal plains from the edges of mangrove swamps to lower upland "bunchgrass/pine savannas", on well-drained, sandy or gravelly, acidic soils. This species forms pure stands or is mixed with *P. oocarpa* and/or *P. tecunumanii*; the vegetation is a fire-climax forest over much of the range of this species.

Distribution: Mexico (S Quintana Roo), Belize, N Guatemala, Honduras (incl. Islas de la Bahía), Nicaragua (where it is the southernmost pine in America).

Altitude: 1–700(–1000?) m.

Notes: Seedlings have an elongated stem and delayed branching.

Related species, subspecies or varieties: In the Caribbean occur two other varieties: *P. caribaea* var. *caribaea* (W Cuba) and *P. caribaea* var. *bahamensis* (Grisebach) W. H. Barrett & Golfari (Bahamas). In Florida (U.S.A.) occurs the related species *P. elliottii*, which was formerly identified as *P. caribaea*. There are no closely related species in Central America, although hybridization with *P. oocarpa* (No. 20) has often been inferred from "intermediate" trees.

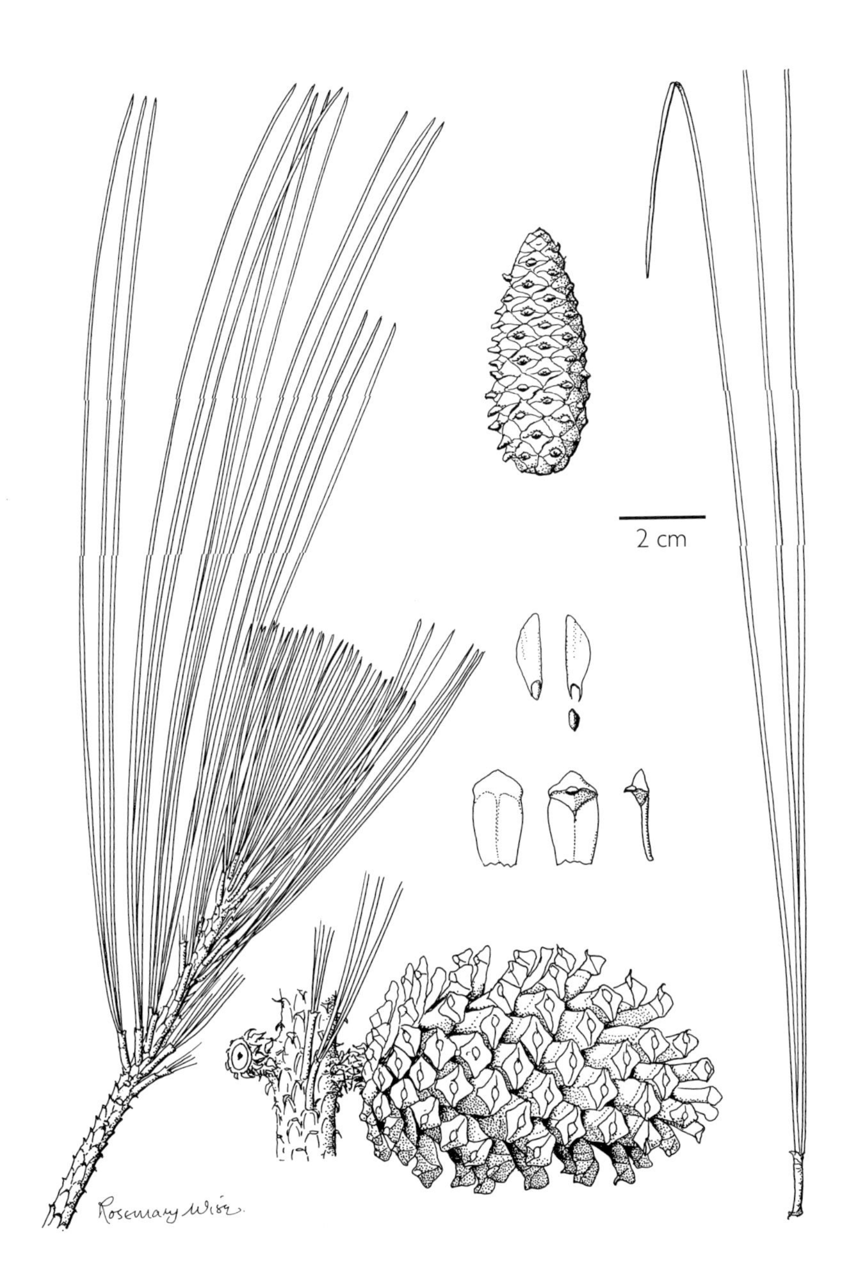
2 cm
Rosemary Wise

9 **Pinus ponderosa** var. **scopulorum**

Pinus ponderosa C. Lawson var. *scopulorum* Engelmann

Local names: Pino

Habit, trunk: tree to 40 m tall, d.b.h. 80–120 cm, trunk straight.

Bark: thick on the trunk, scaly, dividing into large plates separated by broad, shallow fissures, clear orange-brown or reddish-brown with dark fissures.

Foliage twigs: stout, rough with persistent leaf bases (fascicle bases), reddish-brown, often glaucous; needle-fascicles spreading, persisting 2–2.5 years.

Needles: in fascicles of (2–)3 (in some trees predominantly 2 or 3), (10–)15–25(–27) cm long, 1.3–1.6 mm wide, straight or curved, rigid.

Cones: solitary or in whorls of 2–3, nearly sessile, persisting a few years after seed dispersal and leaving a few basal scales when falling, ovoid to subglobose, 5–10 × 4.5–7 cm when open.

Cone scales: 90–120, flexible, opening soon; apophysis raised, transversely keeled; umbo flat or raised, with a persistent, curved spine or prickle.

Seeds: 5–7 × 4–5 mm, light grey-brown, sometimes with dark spots; wing articulate, 15–20 × 6–10 mm, light yellowish-brown, translucent.

Habitat: montane to high montane pine or mixed conifer forest or pine-oak forest.

Distribution: MEXICO: found in two separate localities in Chihuahua (San Luis Range) and Coahuila (Sierra del Carmen). Widely distributed in the U.S.A.

Altitude: not recorded, but above 1500 m.

Notes: *Pinus ponderosa* is a widely distributed species in North America; *P. ponderosa* var. *scopulorum* is the name usually given to the interior (Rocky Mountain) populations. It may occur elsewhere this side of the Mexican border, but not in Baja California (Martínez, 1948), the trees there identified as *P. ponderosa* are all *P. jeffreyi* (No. 11).

Related species, subspecies and varieties: *Pinus arizonica* (No. 10) and its varieties are closely related to *P. ponderosa* and have been classified as a variety of it in more conservative treatments. *Pinus arizonica* mainly differs in its higher number of needles per fascicle (3–5) and its thicker, more rigid cone scales. Other characters are more variable and partly overlap with *P. ponderosa* (see Farjon & Styles, Flora Neotropica Monograph 75).

2 cm
Rosemary Wise

10 Pinus arizonica

Pinus arizonica Engelmann var. *arizonica*

Synonyms: *P. ponderosa* var. *arizonica* (Engelmann) G. R. Shaw

Local names: Pino amarillo, Pino blanco, Pino chino, Pino real (incl. varieties)

Habit, trunk: tree to 30–35 m tall, d.b.h. 100–120 cm, trunk straight.

Bark: thick on the trunk, scaly, breaking into large plates divided by broad, shallow or deep fissures, inner bark brown, outer bark grey.

Foliage twigs: stout, rough with persistent leaf bases (fascicle bases), at first orange-brown or glaucous; needle fascicles spreading or slightly drooping, persisting 2–3 years.

Needles: in fascicles of 3–5, (8–)10–20(–23) cm long, 0.9–1.6 mm wide, straight or slightly curved, rigid to slightly lax.

Cones: in whorls of 2–5, nearly sessile, persisting a few years after seed dispersal, opening soon, (4.5–)5–7 × 3.5–6 cm when open.

Cone scales: 90–140, opening soon, thick, rigid; apophysis raised, transversely keeled, light brown; umbo variable, prickle minute, deciduous.

Seeds: 4–6 × 3–3.5 mm, with an articulate wing 12–15 × 4–6 mm.

Habitat: montane to high montane pine and pine-oak forests, usually on deep, well drained soils; this species is a common tree of both forest types.

Distribution: MEXICO: mainly in the Sierra Madre Occidental south to S Durango, scattered in Coahuila, NE Zacatecas and Nuevo León; also in SW U.S.A.

Altitude: (1300–)200–2700(–3000) m.

Related species, subspecies or varieties: This species is subdivided into three varieties (Farjon & Styles, Flora Neotropica Monograph 75). *P. arizonica* var. *cooperi* (C. E. Blanco) Farjon (syn. *P. cooperi*) has shorter ((5–)6–10(–12) cm) and slightly thinner (1.0–1.3 mm) needles in fascicles of predominantly 5 (but (3–)4–5) and occurs mainly in Durango, but scattered NW into Chihuahua. *Pinus arizonica* var. *stormiae* Martínez differs from var. *arizonica* in its thicker (1.4–1.8 mm), curved and twisted, rigid leaves and its slightly larger cones. It occurs mainly in S Nuevo León, also in S Coahuila, Zacatecas and possibly San Luis Potosí. *Pinus arizonica* is related to *P. ponderosa* (No. 9), of which it is regarded as a variety by conservative botanists.

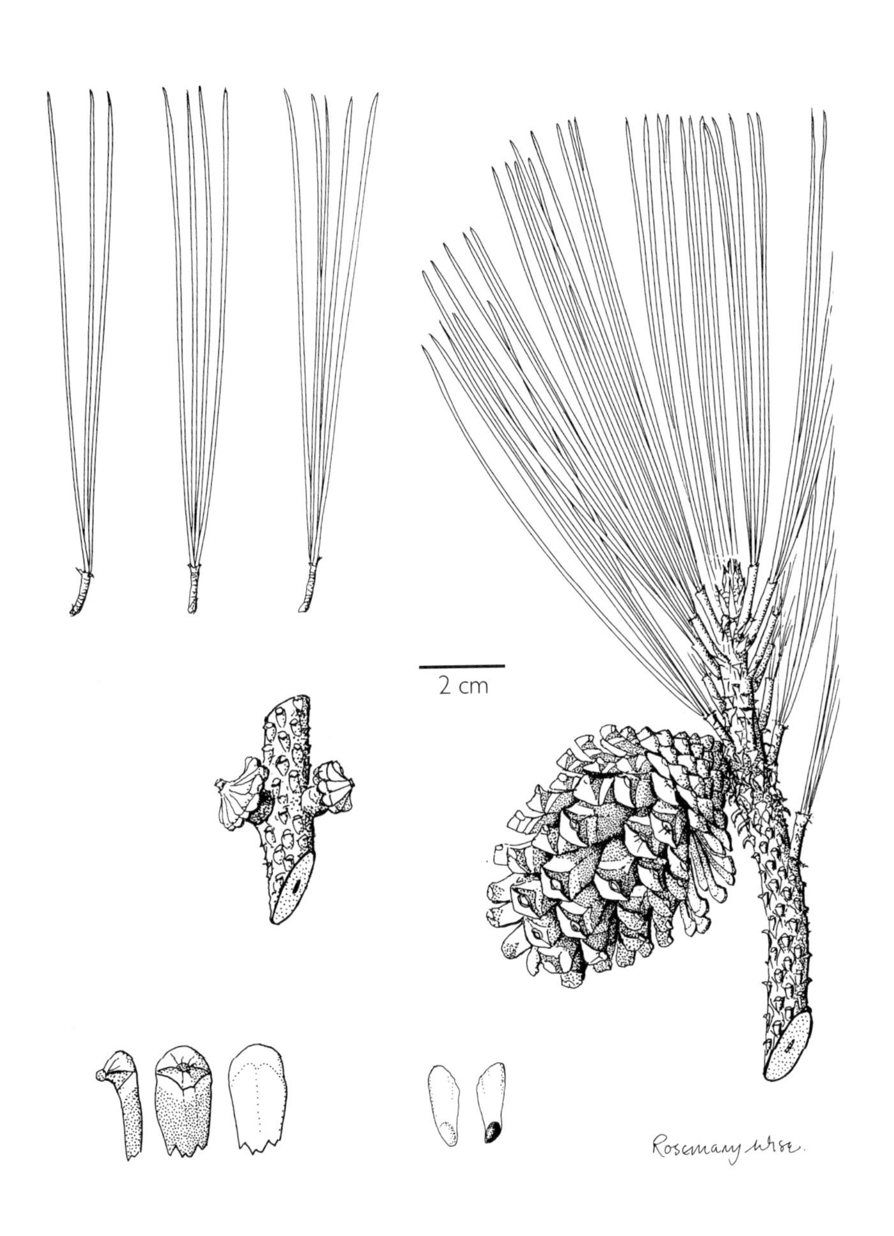
2 cm
Rosemary Wise

11 Pinus jeffreyi

Pinus jeffreyi J. H. Balfour

Local names: Pino, Pino negro

Habit, trunk: tree to 20–30 m tall, d.b.h. to 100 cm, trunk straight.

Bark: thick on the trunk, scaly., with thick, elongated plates divided by deep fissures, light brown, the fissures darker.

Foliage twigs: stout, upturned, very rough with persistent leaf bases (fascicle bases), light orange-brown, often glaucous; needle fascicles spreading, persisting 4–5 years.

Needles: in fascicles of 3, sometimes a few fascicles of 2, (12–)15–22(–25) cm long, 1.5–1.9(–2.0) mm wide, straight or slightly curved, rigid.

Cones: solitary or in pairs, seemingly sessile at maturity, broadly ovoid, base oblique, flattened, (10–)12–17 × 9–14 cm when open, leaving a few basal scales when falling.

Cone scales: 150–175, thin, opening soon and wide; apophysis slightly raised and transversely keeled, often resinous, light brown; umbo with a prominent, persistent prickle.

Seeds: 9–12 mm long, with articulate wing 20–25 × 10 mm, both light brown or yellowish-brown.

Habitat: montane to high montane open pine forest and mixed conifer forest, on deep, well drained soils.

Distribution: MEXICO: Baja California Norte, in the Sierra de Juarez and more common in the northern part of the Sierra San Pedro Martír; it is widespread in the Sierra Nevada of Alta California north into Oregon.

Altitude: Sierra Juarez: (1100–)1400–1800 m; Sierra San Pedro Martír: 1800–2500(–2700) m.

Notes: The above description does not include trees in the U.S.A., where trees in the Sierra Nevada may be much larger and have larger cones and a reddish-brown bark.

Related species: Martínez (1948) confused the trees of this species in Baja California in part with *P. ponderosa* (No. 9), which does not occur there. By some conservative authors, *P. jeffreyi* has been classified as a variety of *P. ponderosa*, but it is now generally accepted as a distinct species.

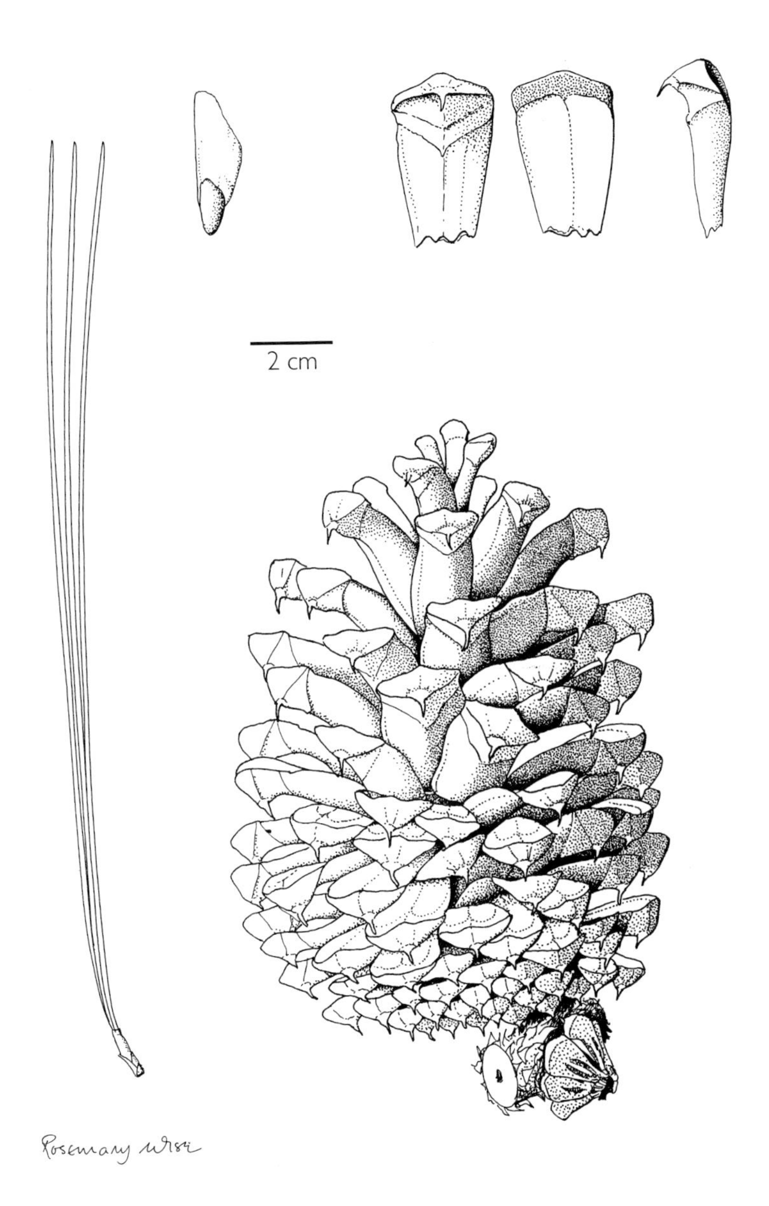
2 cm
Rosemary Wise

12 Pinus engelmannii

Pinus engelmannii Carrière

Local names: Pino real

Habit, trunk: tree to 20–25(–27) m tall, d.b.h. 70–90 cm, trunk straight.

Bark: thick on the trunk, scaly, divided into long, irregular plates by wide, shallow, dark fissures, inner bark brown, outer bark grey.

Foliage twigs: stout, very rough with large, persistent leaf bases (fascicle bases); needle fascicles spreading or slightly drooping, persisting 2–3 years.

Needles: needles in fascicles of (2–)3(–4), rarely 5, (18–)20–35 cm long, 1.5–2.0 mm wide, straight, rigid or slightly lax; fascicle sheaths persistently long, (15–)25–35(–40) mm.

Cones: in whorls of 2–5, seemingly sessile, ovoid-oblong, curved, 8–15 × 6–10 cm when open, leaving a few basal scales when falling.

Cone scales: 100–140, opening gradually, thick; apophysis prominently raised, transversely keeled, often recurved and slightly larger on one side of the cone, light brown; umbo large, with a persistent, curved prickle.

Seeds: 5–8 × 4–5.5 mm, often with dark spots, with articulate wing 18–25 × 7–10 mm, yellowish-brown, translucent.

Habitat: montane to high montane pine and pine-oak forest or more open woodland, on various soils.

Distribution: MEXICO: mainly in the Sierra Madre Occidental in Sonora, Chihuahua, NE Sinaloa, Durango and disjunct in Zacatecas; also extending into the U.S.A. (Arizona, New Mexico).

Altitude: (1200–)1500–2700(–3000) m, most abundantly between 2000–2500 m.

Notes: In the Sierra Madre Occidental it is the only pine with very long leaf fascicles and long, thick leaves.

Related species: *Pinus arizonica* (No. 10) and *P. ponderosa* (No. 9) are related species but these are less robust in their twigs and especially leaf fascicle sheaths, needles and cones.

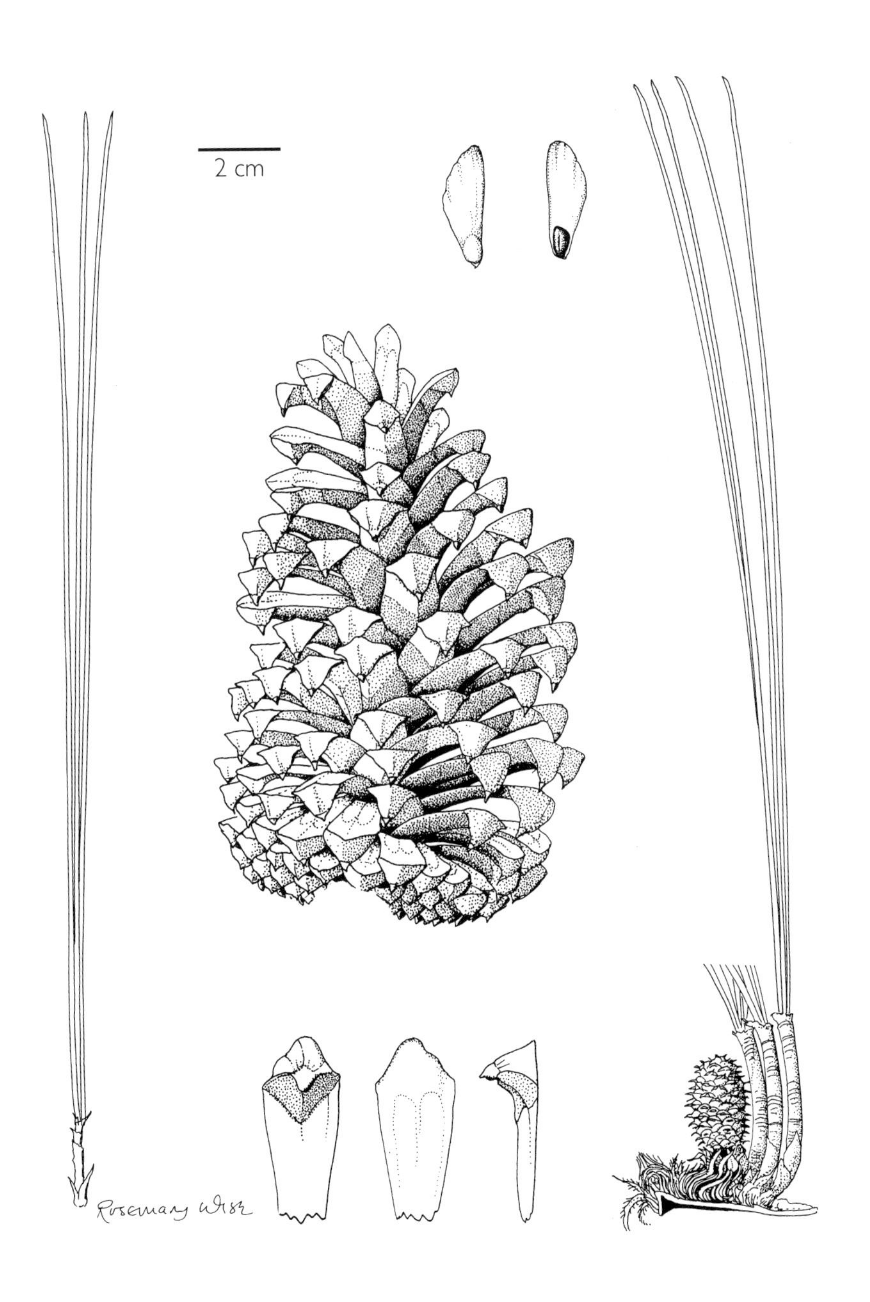
2 cm
Rosemary Wise

13 Pinus hartwegii

Pinus hartwegii Lindley

Synonyms: *P. donnell-smithii* M. T. Masters, *P. rudis* Endlicher

Local names: Ocote, Pino

Habit, trunk: tree to 25–30 m tall, stunted near tree line, d.b.h. 80–100 cm, trunk straight.

Bark: thick on the trunk, very rough and scaly, divided into small or large plates, deeply fissured, dark brown to grey.

Foliage twigs: stout, rigid, upturned, rough with persistent leaf bases (fascicle bases), purplish-brown, sometimes glaucous; needle fascicles densely crowded, spreading, persisting 2(–3) years.

Needles: in fascicles of (3–)4–5(–6), most commonly 5, (6–)10–17(–22) cm long, (1–)1.2–1.5 mm wide, straight or curved, rigid.

Cones: in whorls of 2–6, seemingly sessile, deciduous, obliquely ovoid, 8–12(–14) × 5–8 cm when open.

Cone scales: 150–200, opening soon, thin and flexible or more rigid, spreading widely; apophysis more or less flat, weakly transversely keeled, brown or more often purplish-brown with a blackish, flat or depressed (rarely raised) umbo.

Seeds: 5–6 mm long, often with black spots, with an articulate wing 12–20 × 7–12 mm.

Habitat: This is the true high altitude pine of Mexico and Guatemala, where it often forms extensive pine forests of a single species up to the tree line. In Honduras it is rare, found only on the highest summits and mixed with other conifers. Mixed conifer stands with *P. hartwegii* also occur in Guatemala and Mexico.

Distribution: MEXICO: local in Chihuahua, S Coahuila, S Nuevo León, Durango, SW Tamaulipas, local in Jalisco, Michoacán, Mexico, Morelos, Hidalgo, D.F., Tlaxcala, Puebla, W Veracruz, local in Guerrero, Oaxaca, Chiapas; GUATEMALA highlands, HONDURAS.

Altitude: (2300–)2500–4000(–4300) m.

Notes: In various accounts *P. rudis* is still maintained as a distinct species, but detailed research has demonstrated that pines with characters ascribed to that species grade into pines determined as *P. hartwegii*.

Related species: At its lower limit in Mexico, *P. hartwegii* is often replaced by *P. montezumae* (No. 15), which may be related to it, however these species do not seem to hybridize easily.

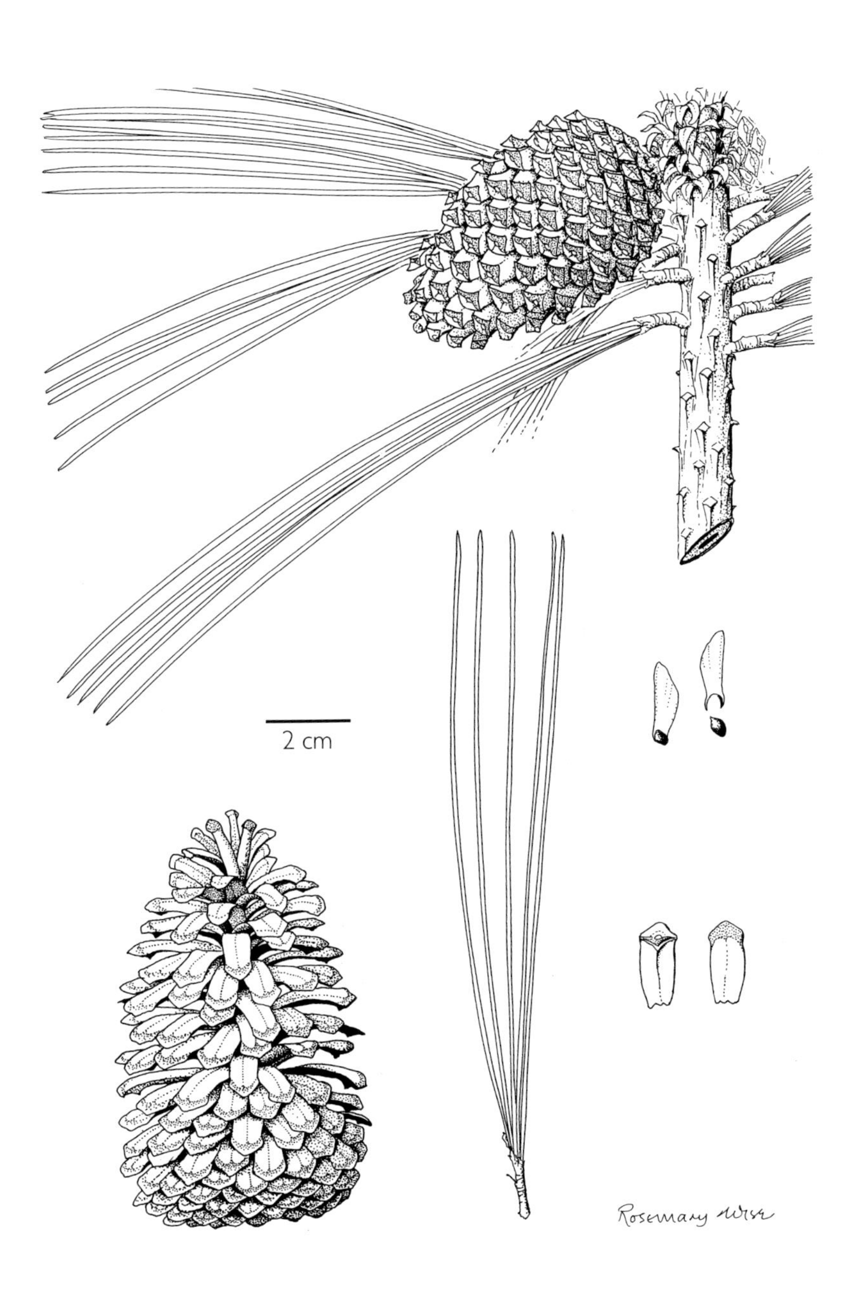
2 cm
Rosemary Wise

14 Pinus pseudostrobus

Pinus pseudostrobus Lindl. var. *pseudostrobus*

Synonyms: *P. pseudostrobus* var. *estevezii* Martínez; *P. pseudostrobus* var. *coatepecensis* Martínez; *P. nubicola* J. P. Perry; (?) *P. yecorensis* Debreczy & Rácz

Local names: Pino blanco, Pino chalmaite, Pino lacio, Pino liso

Habit, trunk: tree to 20–40(–45) m tall, d.b.h. 80–100 cm, trunk straight.

Bark: thick on the trunk, scaly, with elongated plates and deep, longitudinal fissures, dark brown or grey-brown.

Foliage twigs: slender, smooth, with small decurrent leaf bases (fascicle bases), glaucous at first; needle fascicles spreading or more often drooping to nearly pendant; persisting 2–3 years.

Needles: in fascicles of 5, rarely 4 or 6, (18–)20–30(–35) cm long, 0.8–1.3 mm wide, straight, lax, rarely more rigid.

Cones: solitary or in pairs, more rarely in whorls of 3–4 on short, stout peduncles, leaving a few basal scales when falling, 7–16 × 6–13 cm when open, variable, usually asymmetrically ovoid.

Cone scales: 140–190, opening gradually, usually thick woody; apophyses slightly to prominently raised, more so on one side of the cone, transversely keeled, dull brown, weathering greyish; umbo obtuse.

Seeds: 5–7 × 3–4.5 mm, with an articulate wing 20–25 × 7–10 mm covering part of the seed on one side.

Habitat: montane to high montane pine and pine-oak forests; this species is very common and occurs mixed with other pines, or with other conifers.

Distribution: mainly from the Eje Volcanico Transversal (Central México) southward as far as W Honduras, northward there are disjunct populations in Sinaloa/Durango as well as in SE Coahuila/Nuevo León.

Altitude: (850–)1900–3000(–3250) m.

Related species, subspecies or varieties: Due to its variability, a number of species and varieties have been described, of which the only one worthy of consideration is *P. pseudostrobus* var. *apulcensis* (Lindley) Shaw (syn.: *P. oaxacana*, *P. pseudostrobus* var. *oaxacana*). It differs in the strongly elongated apophyses and umbos of the cone scales, but these characters vary and even grade into "typical" *P. pseudostrobus* cones. For detailed information see Farjon (1995) on nomenclature and Farjon & Styles, Flora Neotropica Monopgraph 75.

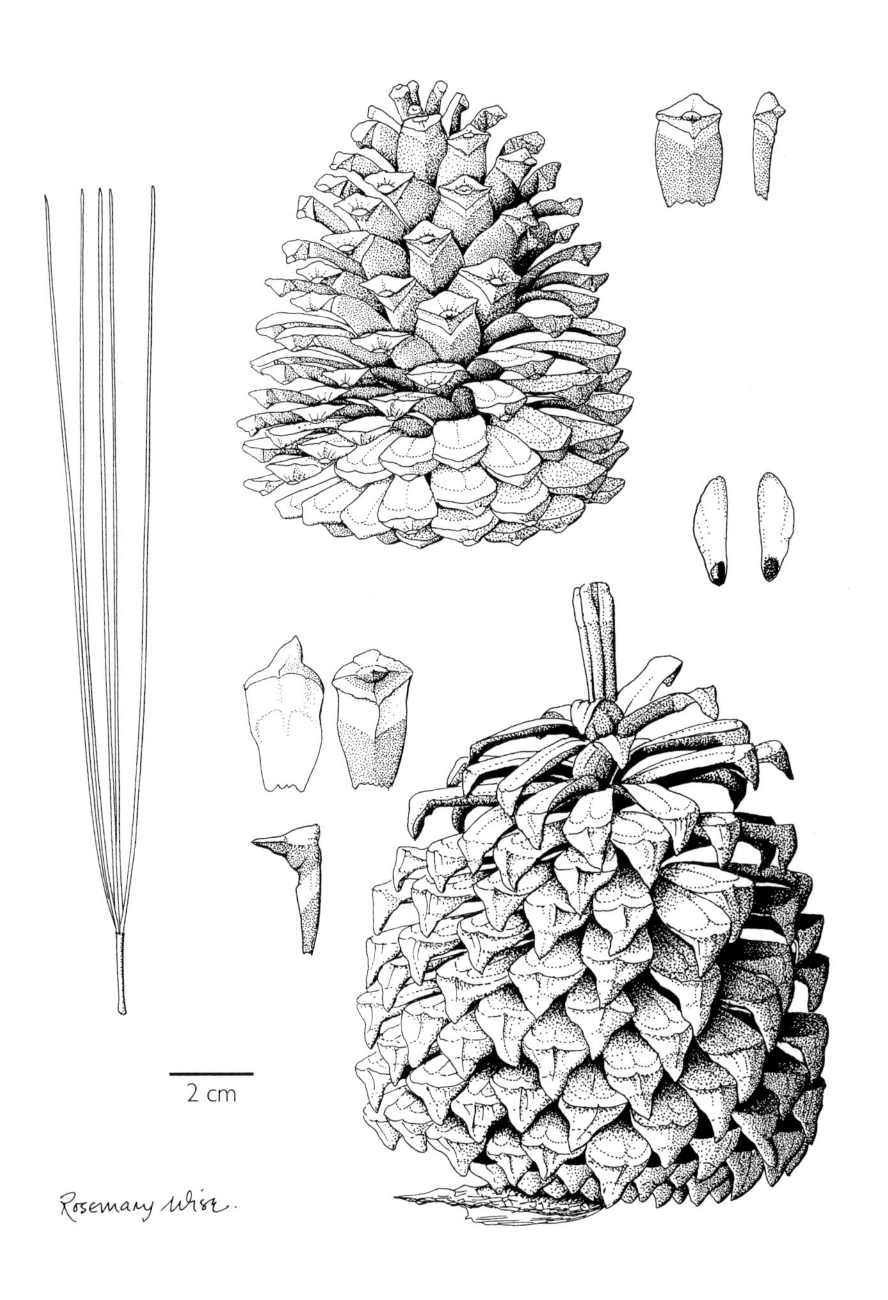
2 cm
Rosemary Wise.

15 Pinus montezumae

Pinus montezumae A. B. Lambert var. *montezumae*

Synonyms: *P. montezumae* var. *lindleyi* J. C. Loudon; *P. montezumae* var. *mezambrana* Carvajal

Local names: Ocote, Ocote blanco, Pino de montezuma, Pino real

Habit, trunk: tree to 20–30 m tall, d.b.h. to 100 cm, trunk straight, often with persistent branches.

Bark: thick on the trunk, scaly, breaking into numerous small, irregular plates divided by shallow fissures, dark brown to grey-black.

Foliage twigs: slender to thick, rough with persistent leaf bases (fascicle bases), brown, new shoots sometimes glaucous; needle fascicles spreading or drooping, persisting 2–3 years.

Needles: in fascicles of (4–)5, rarely 3 or 6, fascicle sheaths (20–)25–35 mm long, 1.5–2.5 mm wide, needles (15–)20–35(–40) cm long, 1.0–1.3 mm wide, straight, lax or more rigid.

Cones: solitary or in whorls of 3–6 on short, stout peduncles, leaving a few basal scales when falling, 8–20 × 5–10 cm when open, variable, usually twice as long as wide, curved.

Cone scales: 175–250, gradually opening, thin or thick woody, rigid; apophysis raised, especially on basal scales, transversely keeled, umbo variable, without a prickle.

Seeds: 5–7 × 4–5 mm, with an articulate wing 18–28 × 7–12 mm, wing and seed both light brown with darker markings.

Habitat: montane to high montane pine and pine-oak forests; this pine is most abundant in the warm temperate zone.

Distribution: MEXICO: in Nuevo León, SW Tamaulipas, Nayarit, S Zacatecas, Jalisco, Michoacán, México, D.F., Querétaro, Hidalgo, Morelos, Tlaxcala, Puebla, Central Veracruz, Guerrero, Oaxaca and Chiapas; GUATEMALA: highlands.

Altitude: (1200–)2000–3200(–3500) m.

Notes: due to its wide distribution this is a variable species, which may hybridize with related species.

Related species, subspecies or varieties: *Pinus montezumae* and *P. devoniana* (No. 16) are mostly separated by continuous characters such as needle and cone size. There are many trees which have "intermediate" needles and/or cones. We keep them separated only because we think that true genetic differences may yet be discovered. A variety: *P. montezumae* var. *gordoniana* (G. Gordon) Silba has slender, lax leaves 0.8–1.0 mm wide and oblong cones with flat, often smooth apophyses.

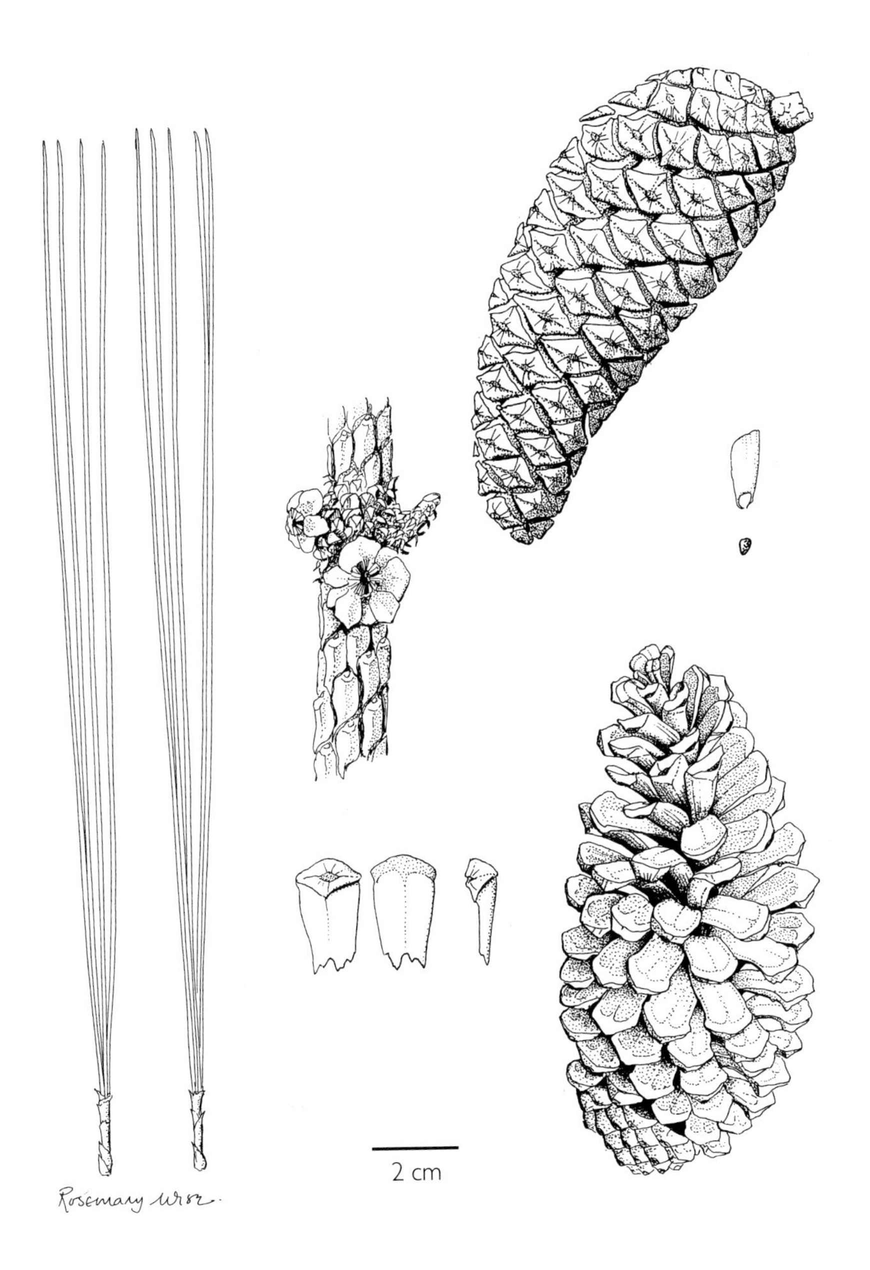
2 cm
Rosemary Wise

16 Pinus devoniana

Pinus devoniana Lindley

Synonyms: *P. michoacana* Martínez; *P. michoacana* var. *cornuta* Martínez; *P. michoacana* var. *quevedoi* Martínez

Local names: Pino blanco, Pino lacio, Pino prieto

Habit, trunk: tree of medium size, sometimes to 20–30 m tall, d.b.h. 80–100 cm, trunk erect, often with persistent but few branches.

Bark: thick on the trunk, scaly, with elongated plates divided by deep fissures, brown, with darker fissures.

Foliage twigs: very thick (15–20 mm), curved, very rough with large leaf bases (fascicle bases); needle fascicles spreading wide, slightly drooping, persisting 2–3 years.

Needles: in fascicles of 5, rarely 4 or 6, fascicle sheaths very long, up to 40 mm, resinous, needles very long(17–)25–40(–45) cm long, 1.1–1.6 mm wide, lustrous green.

Cones: solitary or in whorls of 2–4 on thick, short peduncles, leaving a few scales on the branch when falling, usually large, often curved, 15–35 × 8–15 cm when open.

Cone scales: 175–225, gradually opening, thick woody; apophysis raised and transversely keeled, up to 25 mm wide, umbo flat without a prickle.

Seeds: 8–10 × 5–7 mm, with an articulate wing 25–35 × 10–15 mm.

Habitat: montane, relatively open (secondary) pine or pine-oak forests; this is more a pioneer species than *P. montezumae* and often appears with *P. oocarpa* in disturbed areas.

Distribution: MEXICO: in Nayarit, Jalisco, Zacatecas, Aguascalientes, San Luis Potosi, Querétaro, Hidalgo, Michoacán, México, D.F., Morelos, Tlaxcala, Puebla, Veracruz, Guerrero, Oaxaca and Chiapas; GUATEMALA: in the southern highlands.

Altitude: (700–)900–2500(–3000) m.

Notes: seedlings often develop a "grass stage" with delayed growth of the stem as an adaptation to survive frequent ground fires.

Related species, subspecies or varieties: This species is closely related to *P. montezumae* (No. 15) and also to *P. pseudostrobus* (No. 14) and the first two are sometimes difficult to distinguish. Hybrids probably occur. The cones are especially variable and a number of varieties and forms were described by Martínez, but all grade into each other and appear to have been based on a limited knowledge of this variation in and among populations (see Farjon & Styles, Flora Neotropica Monograph 75). Overall, both foliage and cones are larger in *P. devoniana*.

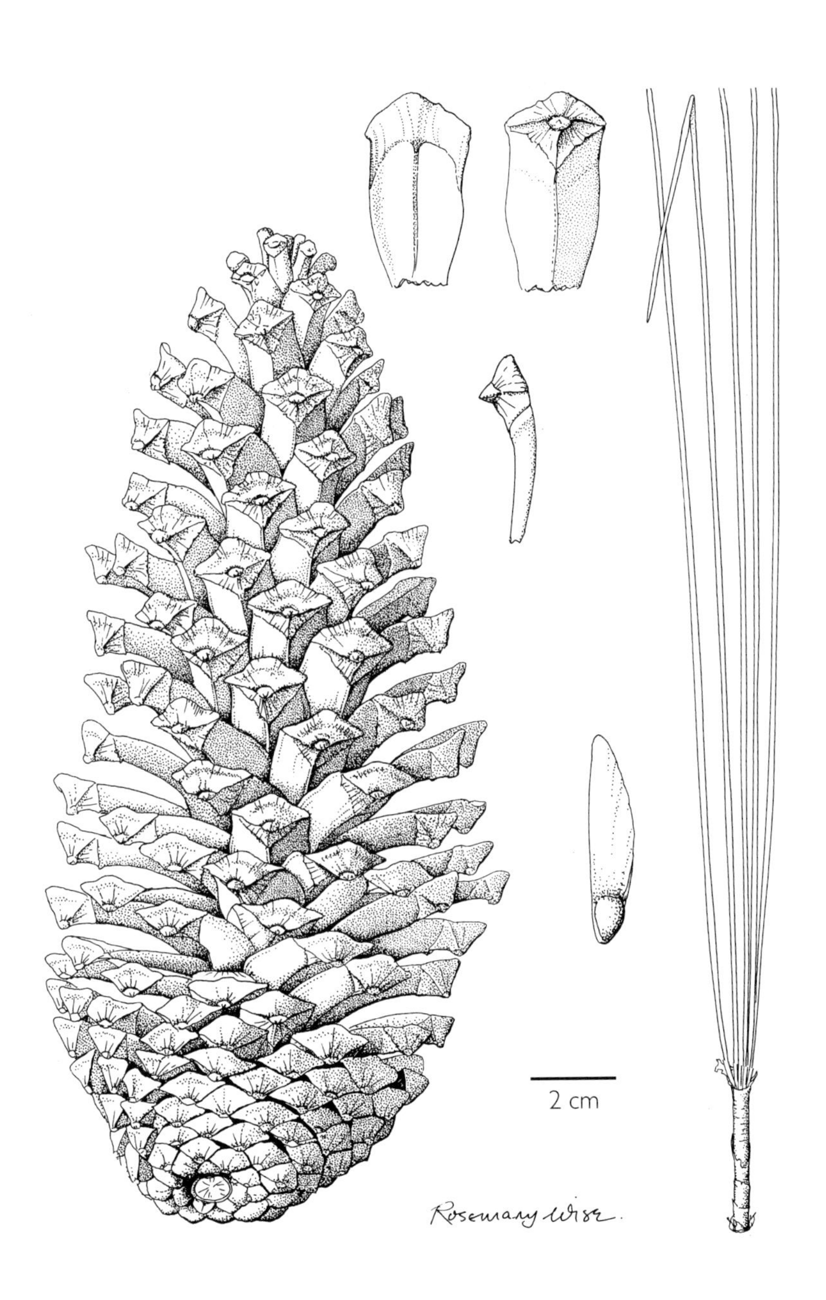
2 cm
Rosemary Wise

17 Pinus douglasiana

Pinus douglasiana Martínez

Local names: Ocote, Pinabete, Pino blanco, Pino hayarín, Pino real

Habit, trunk: tree, to 20–45 m tall, d.b.h. 80–100 cm, trunk straight.

Bark: thick on the trunk, scaly, divided into large, irregular plates and deep fissures, reddish-brown, weathering grey-brown.

Foliage twigs: smooth or rough with prominent but not persistent leaf bases (fascicle bases), dark brown, not glaucous; needle fascicles spreading and drooping, persisting 2–2.5 years.

Needles: in fascicles of 5, rarely 4 or 6, 22–35 cm long, 0.7–1.2 mm wide, straight, lax or sometimes more rigid.

Cones: solitary or in whorls of 3–4 on stout, recurved peduncles which fall with the cone, ovoid-oblong, often slightly curved, 7–10 × 5–7 cm when open.

Cone scales: 110–130, opening soon, rigid, woody; apophysis flat or raised and transversely keeled, with an obtuse umbo.

Seeds: 4–5 × 3–3.5 mm, often with dark spots; wing articulate, 18–24 × 7–9 mm, translucent.

Habitat: montane pine and pine-oak forests in warm temperate to temperate zones.

Distribution: MEXICO: mainly in Jalisco, Michoacán, México, and N Morelos, extending northward to Nayarit and the border between Sinaloa and Durango, also locally in Guerrero and Oaxaca.

Altitude: (1100–)1400–2500(–2700) m.

Notes: A good, but microscopic character is observed when a thin cross-section of a needle, taken from the middle part, is seen under a light microscope (ca. 50x): thick intrusions of hypodermal cells cross over to the central cylinder containing the vascular tissue (see Farjon & Styles, Flora Neotropica Monograph 75; also Martínez, 1948).

Related species: This species is closely related to *P. pseudostrobus* (No. 14), but its cones are smaller and it has the microscopic character in the needles mentioned above to distinguish it. A similar species: *P. maximinoi* (No. 18) has cones with more numerous, very thin, recurving scales.

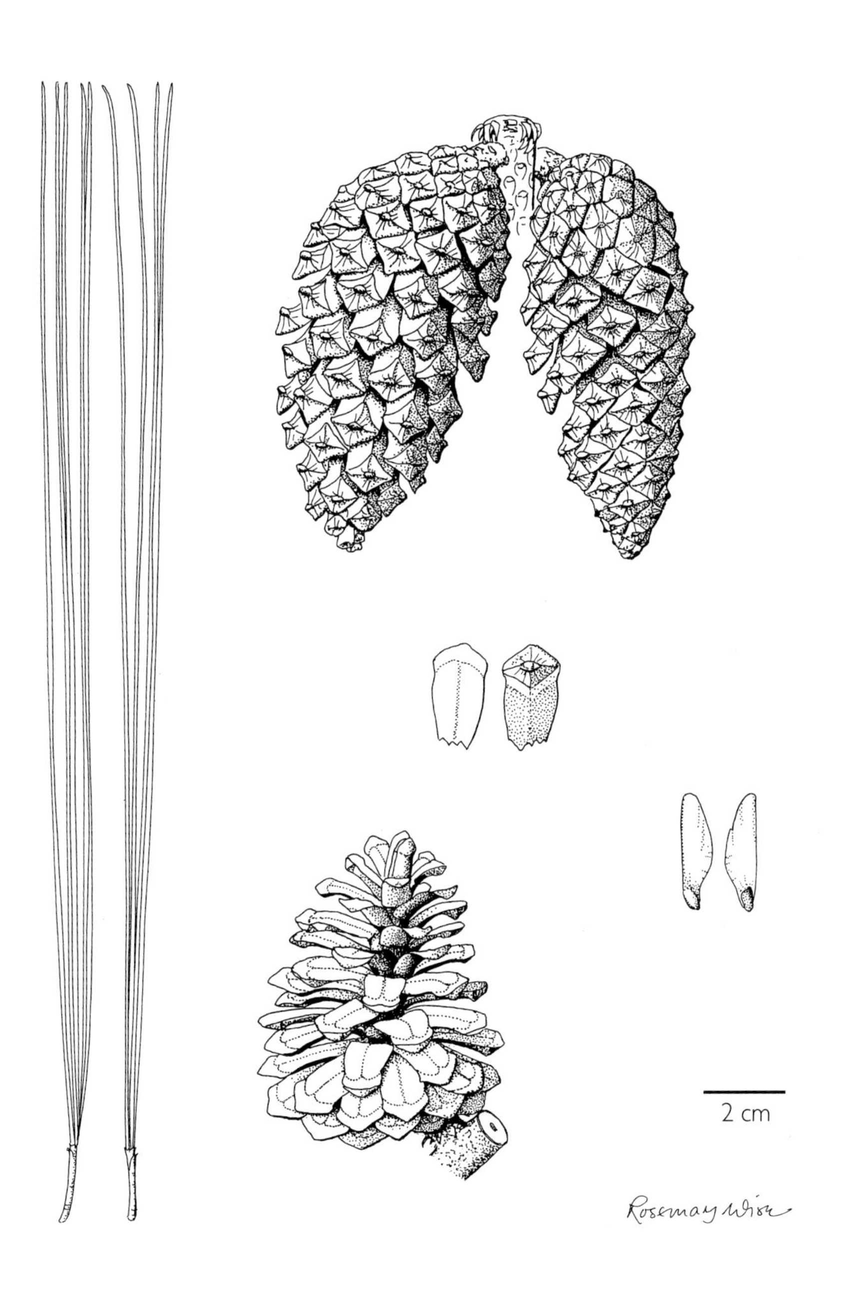
2 cm
Rosemary Wise

18 Pinus maximinoi

Pinus maximinoi H. E. Moore

Synonyms: *P. tenuifolia* Bentham

Local names: Ocote, Pino canís

Habit, trunk: tree to 20–40(–50) m tall, d.b.h. 70–100 cm, trunk straight.

Bark: thick on the lower part of the trunk, with longitudinal plates and fissures, grey-brown.

Foliage twigs: slender, rough with prominent leaf bases (fascicle bases), green or light brown, rarely glaucous; needle fascicles drooping, sometimes pendulous, persisting 2–2.5 years.

Needles: in fascicles of 5, rarely 4 or 6, 20–35 cm long, 0.6–1.0(–1.1) mm wide, lax.

Cones: solitary or in pairs on curved peduncles which fall with the cones, ovoid with an oblique base, (4–)5–10(–12) × (3–)4–8 cm when open.

Cone scales: 120–160, opening soon and usually recurved or reflexed towards the base of the cone, thin and flexible; apophysis flat or slightly raised, light brown, with a raised, darker umbo.

Seeds: 4–6 × 3–4 mm, with an articulate wing 13–22 × 4–8 mm usually lighter coloured than the seed.

Habitat: mainly montane pine and pine-oak forest. This species also occurs as a gap pioneer in wet subtropical forest and cloud forest in Mesoamerica. At lower altitudes it grows in seasonally dry pine woodland with *P. oocarpa*.

Distribution: mainly in the southern half of Mexico, from the Eje Volcánico Transversal through Guatemala, Honduras, El Salvador and NW Nicaragua. A disjunct population is known from Sinaloa in Mexico.

Altitude: (450–)600–2800 m, in the NW of its range 1500–2800 m.

Notes: This species is notable for its long, extremely slender and lax needles, for which it was named *P. tenuifolia* by George Bentham. This name had already been given to a pine from the Himalayas and therefore had to be changed.

Related species: *Pinus maximinoi* has been considered to belong to a "*P. pseudostrobus* group" of closely related pines, in which its nearest relative might be *P. douglasiana* (No. 17). It is distinguished not only by its more slender needles, but also by its thinner, often strongly recurved cone scales.

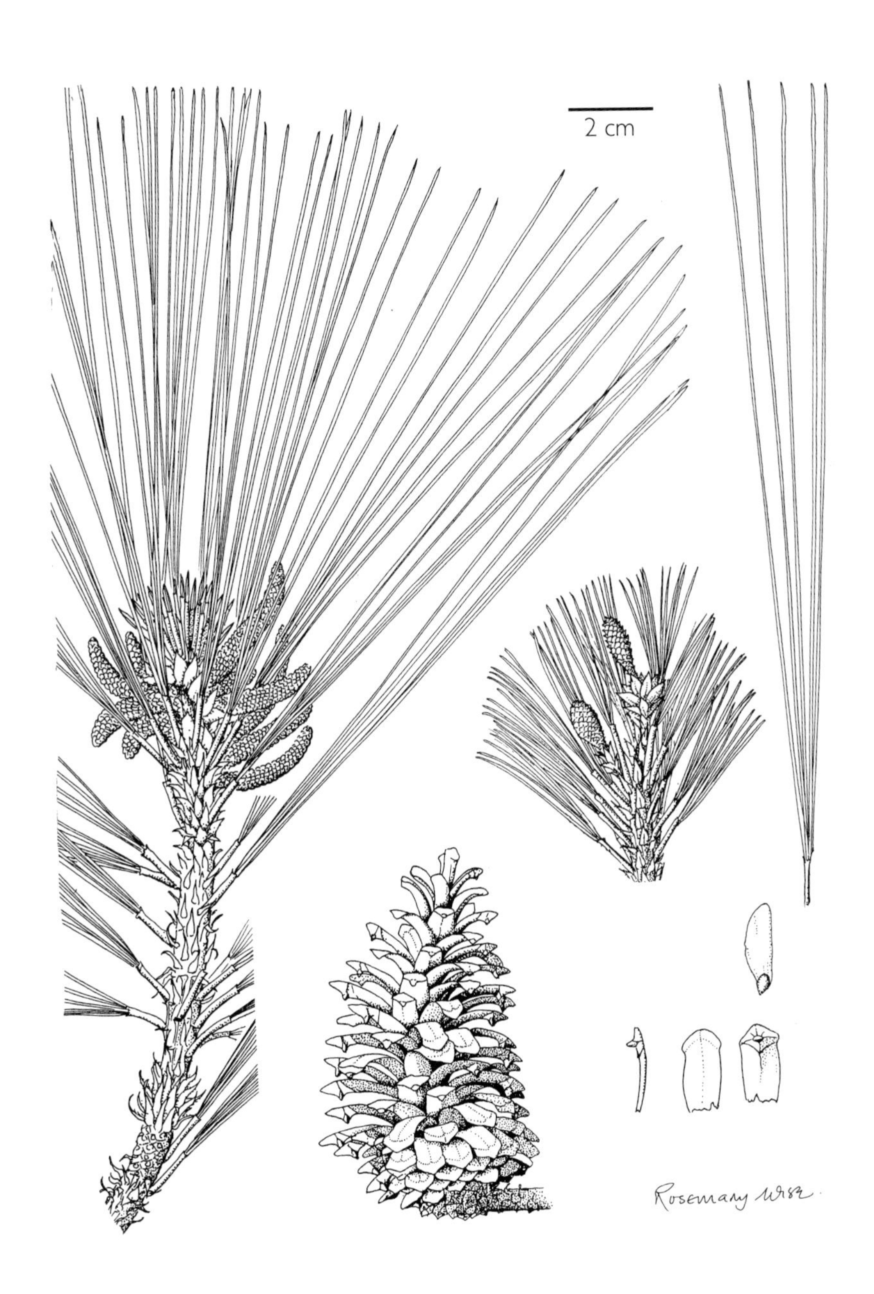
2 cm
Rosemary Wise

19 Pinus lumholtzii

Pinus lumholtzii B. L. Robinson & Fernald

Local names: Ocote dormido, Pino amarillo, Pino barbacaída, Pino lacio, Pino llorón, Pino triste

Habit, trunk: tree to 20 m tall, d.b.h. 50–70 cm, trunk straight.

Bark: thick on the trunk, scaly, divided into elongated, irregular plates and deep longitudinal fissures, outer bark dark grey, nearly black.

Foliage twigs: ridged and grooved, glaucous to reddish-brown, later grey; needle fascicles remote, very pendulous, persisting 2 years.

Needles: in fascicles of 3 (exceptionally 2 or 4), (15–)20–30(–40) cm long, (1–)1.2–1.5 mm wide, lax, light green. Fascicle sheaths 25–35 mm long on juvenile fascicles, soon disintegrating and falling off.

Cones: solitary (sometimes in whorls of 2, rarely 3) on curved peduncles, soon deciduous, (3–)3.5–5.5(–7) × (2.5–)3–4.5 cm when open, broadest near the base.

Cone scales: 70–90, opening gradually, thick woody; apophysis slightly raised and thickened especially near the base of the cone, with an obtuse, darker umbo.

Seeds: 3–5 mm long, dark brown, often with black dots, with an articulate wing 10–14 × 4–6 mm, lighter coloured than the seed.

Habitat: most commonly montane pine-oak forests, also mixed pine forests on wetter slopes of the Sierra Madre Occidental and other mountains. Sometimes this pine occurs in rather dry types of forest on poor soil with *P. cembroides*.

Distribution: MEXICO: mainly in the Sierra Madre Occidental, in Chihuahua, Sinaloa, Durango, Nayarit, Jalisco, Zacatecas, Aguascalientes and Guanajuato.

Altitude: (1500–)1700–2600(–2900) m.

Notes: The extremely pendulous foliage, together with the deciduous fascicle sheaths coming off in long, brown scales, easily distinguish this species.

Related or similar species: The only other species of the "hard pines" with deciduous fascicle sheaths is *P. leiophylla* (No. 1), but this species differs in its much shorter, not pendulous leaves and in characters of its cones (and others) which seem to indicate that these two species are not closely related (see Farjon & Styles, Flora Neotropica Monograph 75).

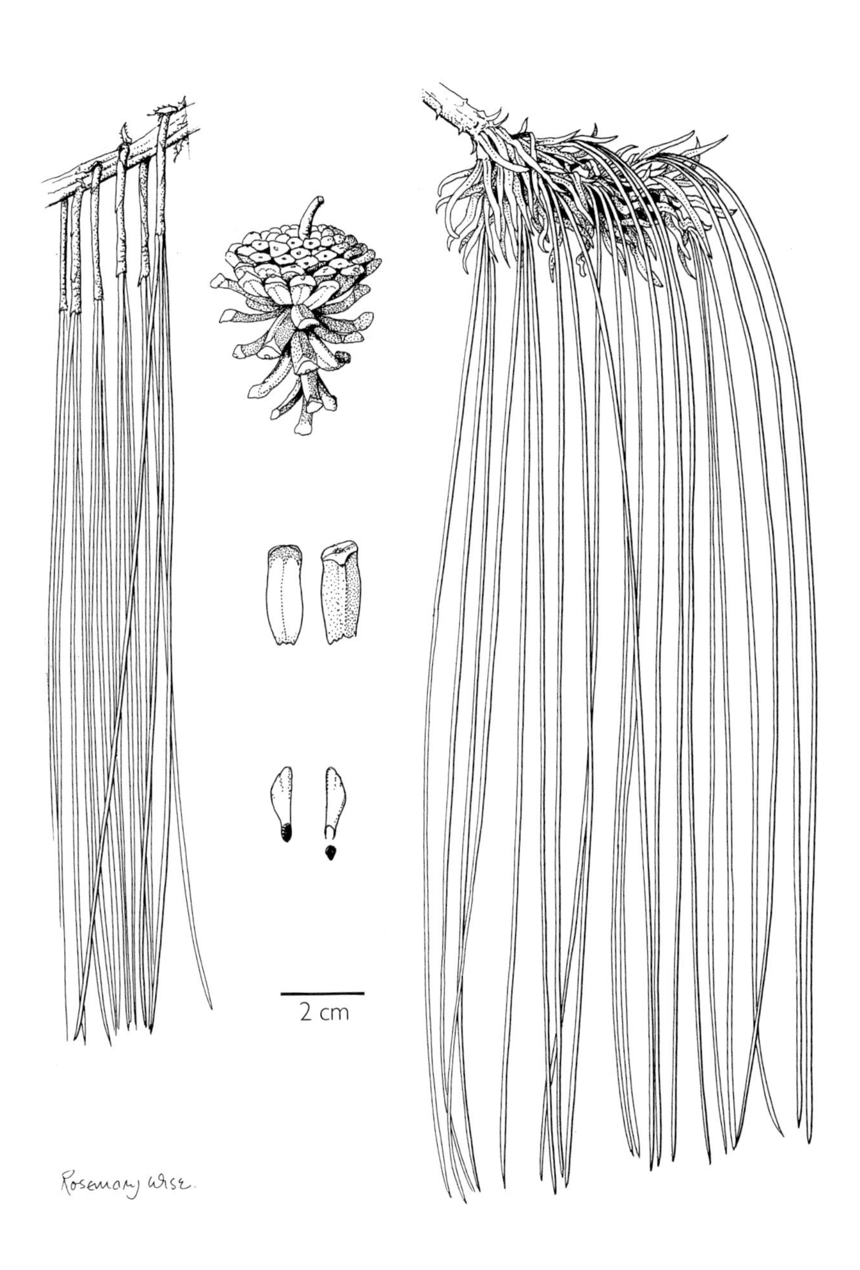
2 cm
Rosemary Wise.

20 Pinus oocarpa

Pinus oocarpa Schlechtendal var. *oocarpa*

Synonyms: *P. oocarpa* var. *manzanoi* Martínez

Local names: Ocote chino, Pino chino, Pino colorado, Pino prieto, Pino tepo

Habit, trunk: tree to 30–35 m tall, d.b.h. to 100–125 cm, trunk straight or tortuous.

Bark: thick on the trunk, scaly, breaking into small or large, longitudinal plates and shallow fissures, reddish-brown to grey-brown.

Foliage twigs: rough with decurrent leaf bases (fascicle bases), reddish-brown; needle fascicles spreading or slightly drooping, persisting 2–3 years.

Needles: in fascicles of 5 (sometimes 3–4 on trees with mostly 5), (17–)20–25(–30) cm long, 0.8–1.4 mm wide, straight, mostly rigid.

Cones: solitary or in whorls of 3–4 on stout, curved peduncles, persisting for several years, broadly ovoid to subglobose when closed, with a flattened base and 3–8(–10) × 3–9(–12) cm when open.

Cone scales: 70–130, opening slowly from the base of the cone, thick woody; apophysis nearly flat or slightly raised especially on the basal scales, light brown, with an obtuse umbo.

Seeds: 4–8 × 3–4.5 mm, grey, often with black spots; wing articulate, 8–18 × 4–8 mm, greyish-brown.

Habitat: a great variety of forest types. In Mesoamerica *P. oocarpa* is often the single species in widely distributed open pine forest, in Mexico it occurs also in open pine and pine-oak forest, from lowland hills to major mountain ranges. Fires are frequent in forests where *P. oocarpa* predominates.

Distribution: from NW Mexico (Sierra Madre Occidental) to Guatemala, Honduras, El Salvador and NW Nicaragua.

Altitude: (200–)500–2300(–2700) m.

Notes: *Pinus oocarpa* usually retains its cones several years after the seeds have been shed, consequently the trees often appear to be full of cones.

Related species, subspecies and varieties: A species with such a large range naturally varies. One botanical variety, which has consistently only 3 leaves in its fascicles and grows on poor, shallow soils (remaining a small tree) is recognized here: *P. oocarpa* var. *trifoliata* Martínez.

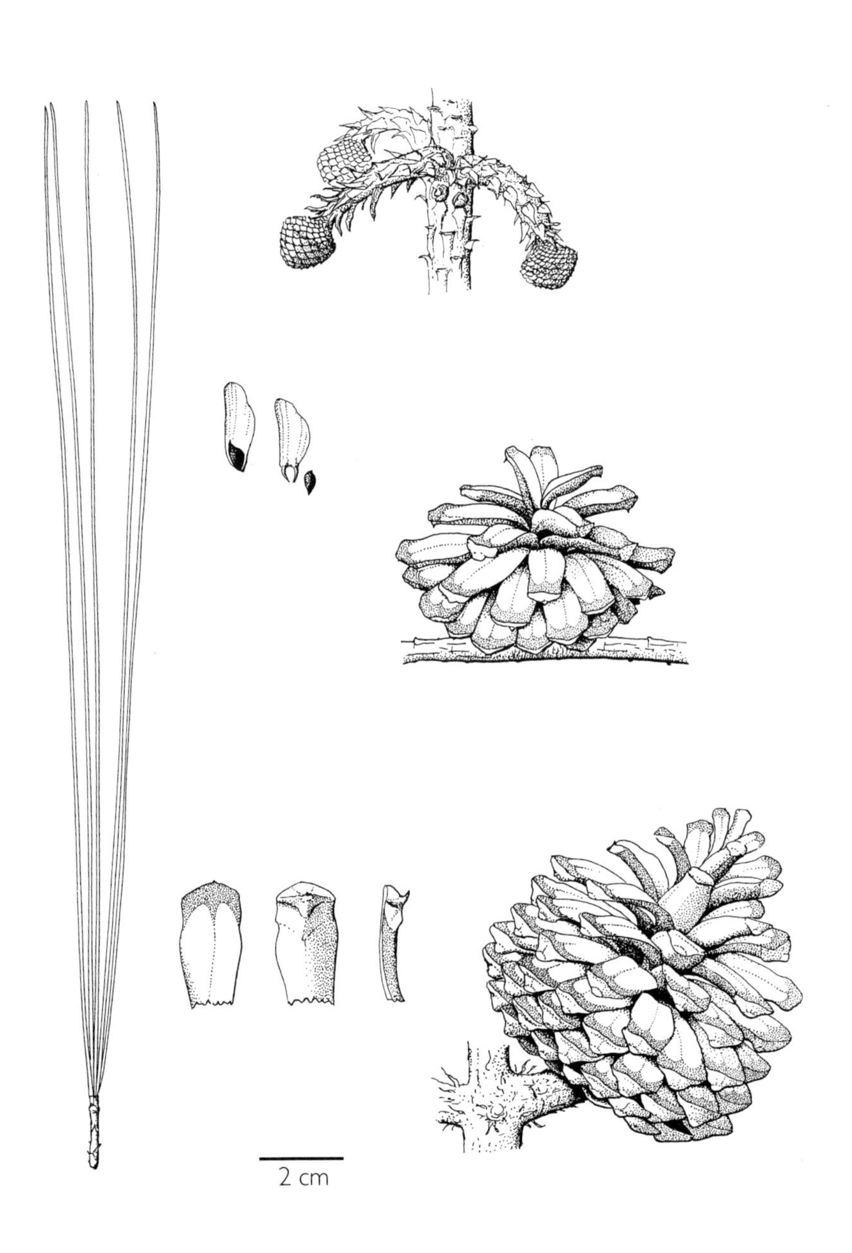
2 cm

21 Pinus praetermissa

Pinus praetermissa Styles & McVaugh

Synonyms: *P. oocarpa* var. *microphylla* G. R. Shaw

Local names: Pino chino, Pino prieto

Habit, trunk: tree to 10–15(–20) m tall, d.b.h. to 30 cm, trunk curved or tortuous, sometimes branching low.

Bark: relatively thin, rough and scaly, breaking in thin, irregular plates and longitudinal fissures, grey-brown.

Foliage twigs: slender, smooth, with small, early eroding leaf bases (fascicle bases), soon scaly and flaky, reddish-brown to greyish; needle fascicles spreading, persisting 3 years.

Needles: in fascicles of (4–)5, (8–)10–16 cm long, 0.5–0.8 mm wide, delicate, lax.

Cones: commonly solitary, sometimes in pairs, on slender, recurved peduncles, deciduous, opening soon, broadly ovoid to subglobose when closed, (4–)5–6.5(–7) × (5–)6–8 cm when open, often loosing the basal scales when still on the tree.

Cone scales: 100–120, opening soon, thin woody but rigid; apophysis flat to slightly raised, lustrous light brown, with a flat or slightly raised, obtuse umbo.

Seeds: 5–8 × 3–4 mm, blackish-grey, with an articulate wing 12–18 × 5–8 mm.

Habitat: open, dry pine-oak woodlands, often on rocky slopes, also tropical broad-leaved forests on poor sites.

Distribution: MEXICO: in S Sinaloa, Nayarit and Jalisco; probably incompletely known.

Altitude: 900–1900 m.

Notes: This species is relatively unknown, probably because it is commonly confused with *P. oocarpa* (No. 20) and therefore not collected. More critical collecting may result in a better knowledge of its distribution and habitat.

Related species, subspecies or varieties: *Pinus praetermissa* has been described as a variety of *P. oocarpa*, mainly because of its cones which resemble it very much in shape. It is, however, quite distinct (see Farjon & Styles, Flora Neotropica Monograph 75). Good field characters are the very slender needles, the fact that the cones fall early and the often missing basal scales on fallen cones.

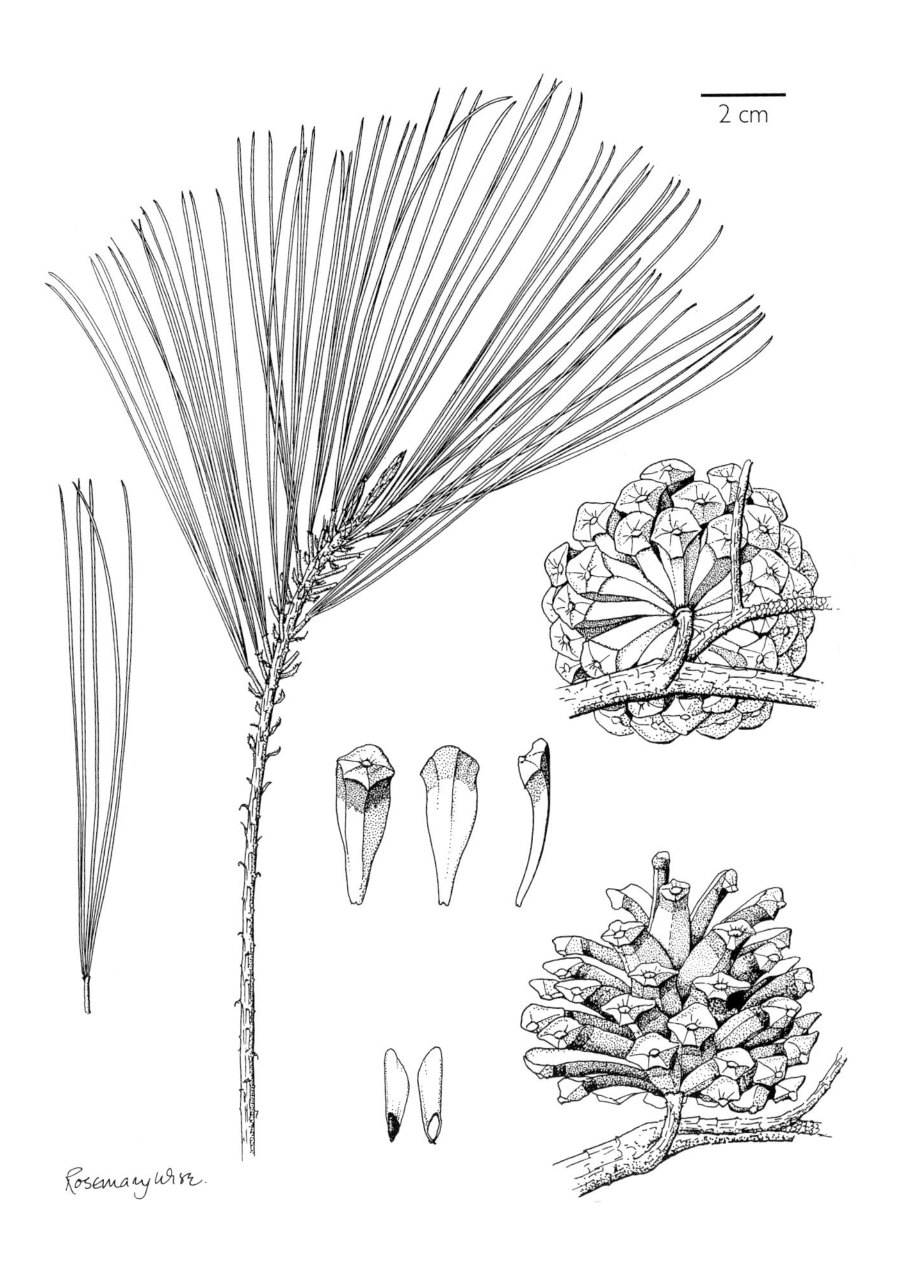
2 cm
Rosemary Wise.

22 Pinus patula

Pinus patula Schlechtendal & Chamisso var. *patula*

Local names: Ocote, Peinador de neblinas, Pino, Pino colorado, Pino lacio, Pino triste

Habit, trunk: tree to 35–40 m tall, d.b.h. 90–100 cm, trunk straight.

Bark: thick on the trunk, scaly, with large, elongated plates which exfoliate easily and deep, longitudinal fissures, dark grey-brown on the lower part of the trunk, thin scaly and reddish-brown in the upper part and on branches.

Foliage twigs: often multinodal, with prominent leaf bases (fascicle bases), yellowish- to reddish-brown, soon scaly; needle fascicles drooping on two sides of the shoot, persisting 2–3 years.

Needles: in fascicles of 3–4(–5), (11–)15–25(–30) cm long, 0.7–0.9(–1) mm wide, straight, lax.

Cones: solitary or in whorls of 2 or more, nearly sessile, persistent, narrowly ovoid or oblong when closed, asymmetrical at base, 5–10(–12) × (3–)4–6.5 cm when open, yellowish-brown, turning grey with age.

Cone scales: 100–150, remaining partly closed or opening slowly, rigid; apophysis nearly flat to slightly raised, more raised on the basal scales, yellowish-brown, with a usually flat umbo.

Seeds: 4–6 × 2–4 mm, dark, with a lighter coloured, articulate wing 12–18 × 5–8 mm.

Habitat: the more humid, subtropical to warm-temperate montane sites in mixed pine or pine-oak forest; on the Atlantic slope this species is also associated with *Liquidambar* and other broad-leaved trees.

Distribution: MEXICO: in Tamaulipas, Querétaro, Hidalgo, México, D.F., Morelos, Tlaxcala, Puebla, Veracruz, Oaxaca and Chiapas.

Altitude: (1400–)1800–2800(–3300) m.

Related species, subspecies or varieties: *Pinus patula* is related to *P. oocarpa* (No. 20), *P. tecunumanii* (No. 24) and *P. jaliscana* (No. 23), some of which have been classified as subspecies or varieties of it. We treat these pines as distinct species and accept *P. patula* var. *longipedunculata* Martínez as a variety. It has distinct peduncles and the cones are not clustered, while they fall after a few years from the branches.

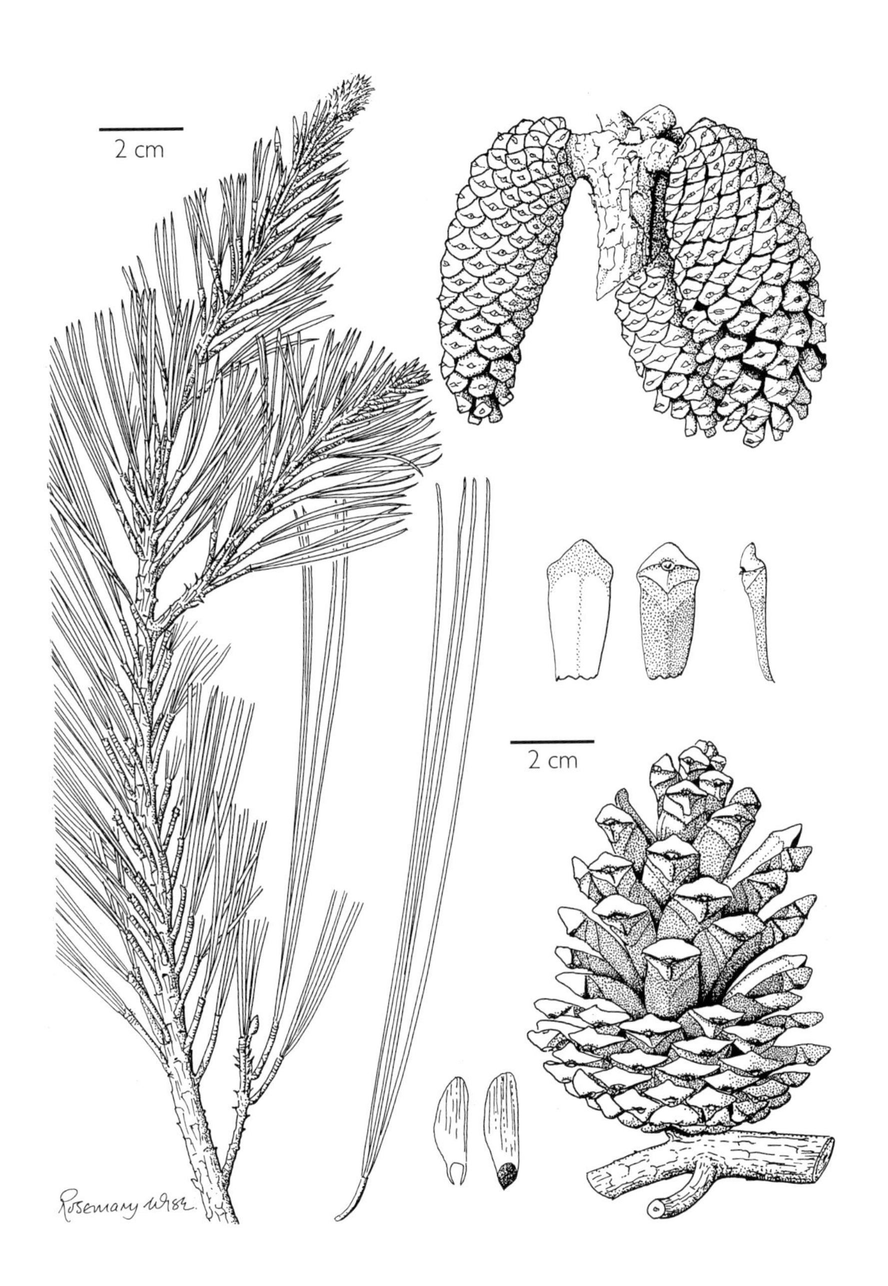
2 cm
2 cm
Rosemary Wise 1982

23 Pinus jaliscana

Pinus jaliscana Pérez de la Rosa

Synonyms: *P. macvaughii* Carvajal

Local names: Ocote, Pino

Habit, trunk: tree to 25–35 m tall, d.b.h. 60–100 cm, trunk straight.

Bark: thick on the lower part of the trunk, scaly, with irregular, elongated plates and shallow fissures, inner bark reddish-brown, outer bark grey-brown.

Foliage twigs: smooth, with small, soon eroding leaf bases (fascicle bases) fand small scales; needle fascicles spreading or slightly drooping, persisting 2–3 years.

Needles: in fascicles of (4–)5, rarely 3, 12–18(–22) cm long, (0.5–)0.6–0.8 mm wide, straight, lax, light green.

Cones: solitary or in whorls of 2–3 on curved peduncles, deciduous, ovoid-oblong, often with an oblique base when closed, (4.5–)6–8.5(–9.8) × (3–)4–5(–6) cm when open.

Cone scales: 90–115, opening gradually but often remaining closed near the base of the cone; apophysis slightly raised, often thicker near the base on one side of the cone, light brown; umbo flat or obtuse, light brown or grey.

Seeds: 3.5–6 × 2–3.5 mm, with an articulate wing 13–17 × 6–8 mm, lighter coloured than the seed.

Habitat: low montane pine or pine-oak forest with *P. maximinoi* and/or *P. oocarpa*, on acidic soils.

Distribution: MEXICO: Jalisco, mainly in the Sierra de Cuale (Sierra de El Tuito) and some neighboring mountains in the NW part of the Sierra Madre del Sur; still incompletely known.

Altitude: 800–1200(–1650) m.

Notes: Where *P. jaliscana* grows together with *P. oocarpa* or *P. maximinoi*, it can be distinguished at some distance by the light green foliage. The needles have large septal resin ducts (cross-section, microscope!).

Related or similar species: This species has only recently been recognized and was at first confused with *P. herrerae* (No. 4), which has needles in fascicles of 3 and very small cones (maximum 4 cm long). It seems to be related to *P. patula* (No. 22), which has drooping, usually longer needles and persistent, clustering cones, and to *P. tecunumanii* (No. 24), which is a Central American pine reaching Oaxaca. That species has wider, more rigid needles and cones resembling those of *P. oocarpa* (No. 20).

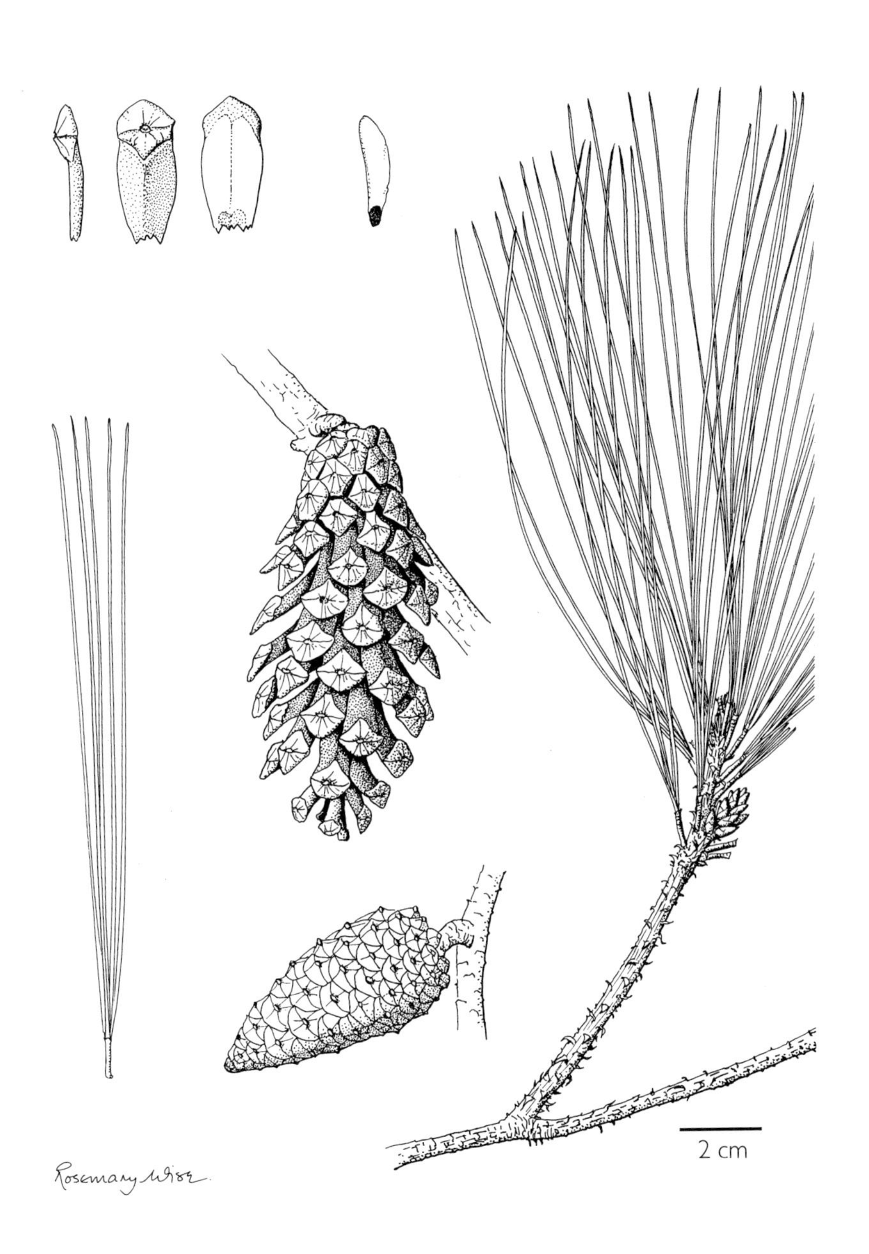
2 cm

24 Pinus tecunumanii

Pinus tecunumanii Eguiluz & J. P. Perry

Synonyms: *P. patula* subsp. *tecunumanii* (Eguiluz & J. P. Perry) Styles; *P. oocarpa* var. *ochoterenae* Martínez

Local names: Ocote, Pino

Habit, trunk: tree to 50–55 m tall, d.b.h. to 120–140 cm, trunk straight, large trees with a clear bole.

Bark: thick on the lower part of the trunk, soon thinning out, with irregular small flakes, dark grey-brown on the lower part of the trunk, reddish-brown above.

Foliage twigs: slender, rough with persistent and decurrent leaf bases (fascicle bases), reddish-brown, often glaucous; needle fascicles drooping, persisting 2–3 years.

Needles: in fascicles of 4(3–5), (14–)16–18(–25) cm long, 0.7–1.0(–1.3) mm wide, straight, lax.

Cones: in whorls of 2–4, rarely solitary, on long, curved peduncles, falling after 1–3 years with the peduncles, ovoid, with a rounded base in open cones, (3.5–)4–7(–7.5) × (3–)3.5–6 cm when open, light brown.

Cone scales: 100–140, opening gradually from the base of the cone to the apex; apophysis raised, transversely keeled, with an obtuse umbo.

Seeds: 4–7 × 2–4 mm, dark grey or blackish, with an articulate wing 10–13 × 4–8 mm, grey-brown.

Habitat: open to closed-canopy pine and pine-oak forests from foothills to montane zones, in regions with abundant rainfall; also gaps of broad-leaved forest with *Liquidambar* and other species.

Distribution: MEXICO: Oaxaca, Chiapas, possibly SE Guerrero; more widespread in Guatemala, Belize, Honduras, El Salvador and Nicaragua.

Altitude: (300–)550–2500(–2900) m.

Related species, subspecies and varieties: *Pinus tecunumanii* is closely related to *P. patula* (No. 22) and *P. oocarpa* (No. 20). In Central America it is often regarded as a distinct species, while in Mexico what is almost certainly the same species has been classified as *P. oocarpa* var. *ochoterenae* by Martínez. The cones of *P. tecunumanii* resemble those of *P. oocarpa*, but have a slender peduncle, are not wider than long and have a rounded base when open. They fall earlier from the tree than in the other species. Pinus patula has ovoid-oblong cones, usually nearly sessile when mature. For more information see Farjon & Styles, Flora Neotropica Monograph 75.

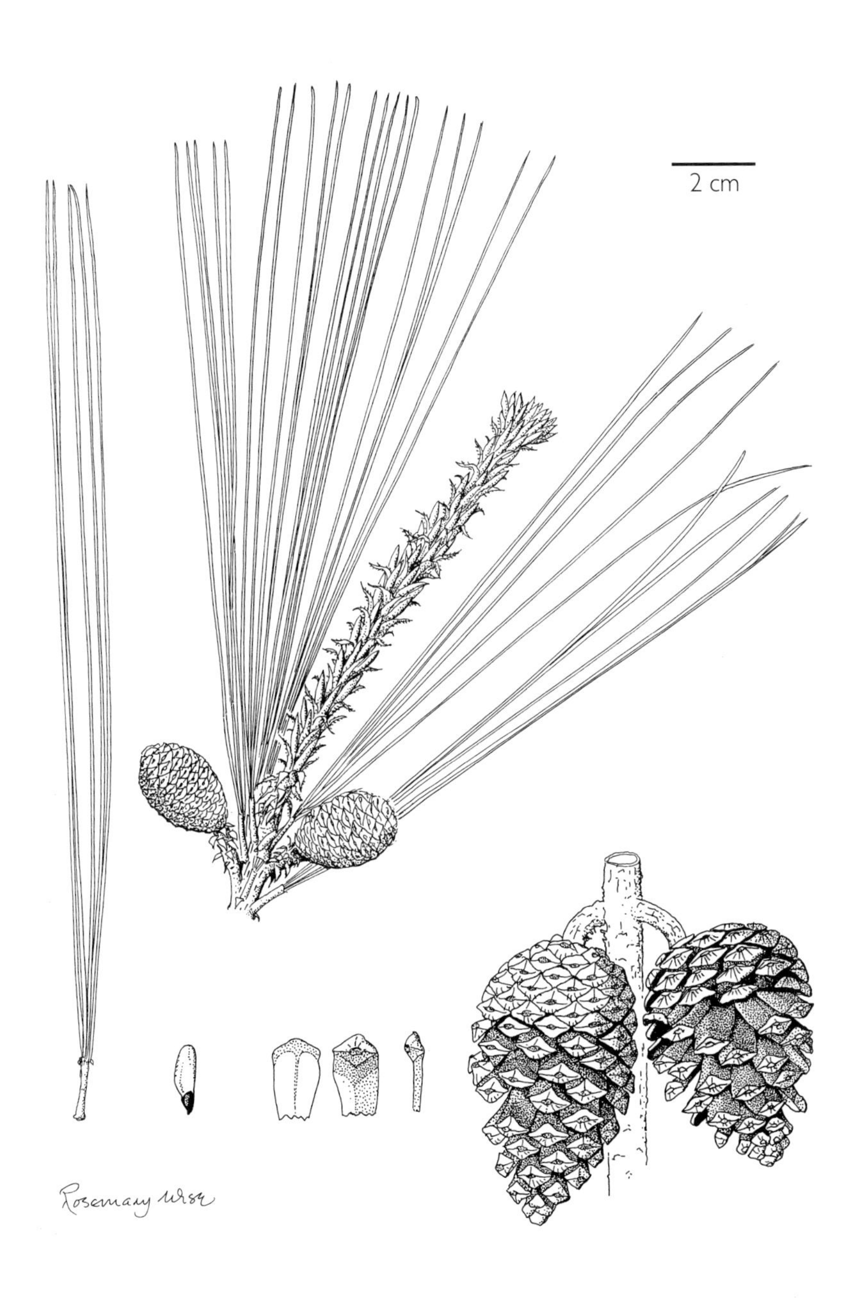
2 cm
Rosemary Wise

㉕ Pinus durangensis

Pinus durangensis Martínez

Synonyms: *P. martinezii* E. Larsen

Local names: Ocote, Pino blanco, Pino real

Habit, trunk: tree to 35–40 m tall, d.b.h. 80–100 cm, trunk straight.

Bark: thick on the trunk, scaly, breaking into large, irregular, elongated plates and shallow fissures, brown, turning grey.

Foliage twigs: rough with persistent leaf bases (fascicle bases), orange-brown or reddish-brown, usually glaucous; needle fascicles spreading or drooping, persisting 2–2.5 years.

Needles: in fascicles of (4–)5–6(–7, rarely 8), 14–24 cm long, 0.7–1.1 mm wide, straight or slightly curved, lax or more rigid.

Cones: Solitary or in whorls of 2–4 on short peduncles, only falling after several years, 5–9(–11) × 4–6(–7) cm and ovoid with a flattened base when completely open, yellowish-brown to brown.

Cone scales: 90–120, thick woody, opening gradually; apophysis flat to raised, transversely keeled; umbo raised, slightly recurved, with a small prickle.

Seeds: 5–6 × 4–4.5 mm, with an articulate wing 14–20 × 6–9 mm.

Habitat: montane pine and pine-oak forest, on shallow as well as deep soils; this species grows in pure stands or mixed with other species.

Distribution: MEXICO: mainly in the southern part of the Sierra Madre Occidental, rare in E Sonora and Chihuahua, common in Durango, Zacatecas and N Jalisco, more scattered further south in Jalisco and N Michoacán.

Altitude: (1400–)1600–2800(–3000?) m.

Notes: Trees of this species sometimes have more than 6 needles in a fascicle. The colour of the leaves is variable from yellowish-green to glaucous-green.

Related or similar species: *Pinus arizonica* (No. 10), which also occurs in the Sierra Madre Occidental, has similar cones but thicker and more rigid needles in fascicles of 3–5. *Pinus lawsonii* (No. 26) and *P. pringlei* (No. 27) are closely related, but occur only in southern Mexico. *Pinus teocote* has needles in fascicles of 3 and cone scales with flat, not raised, umbos.

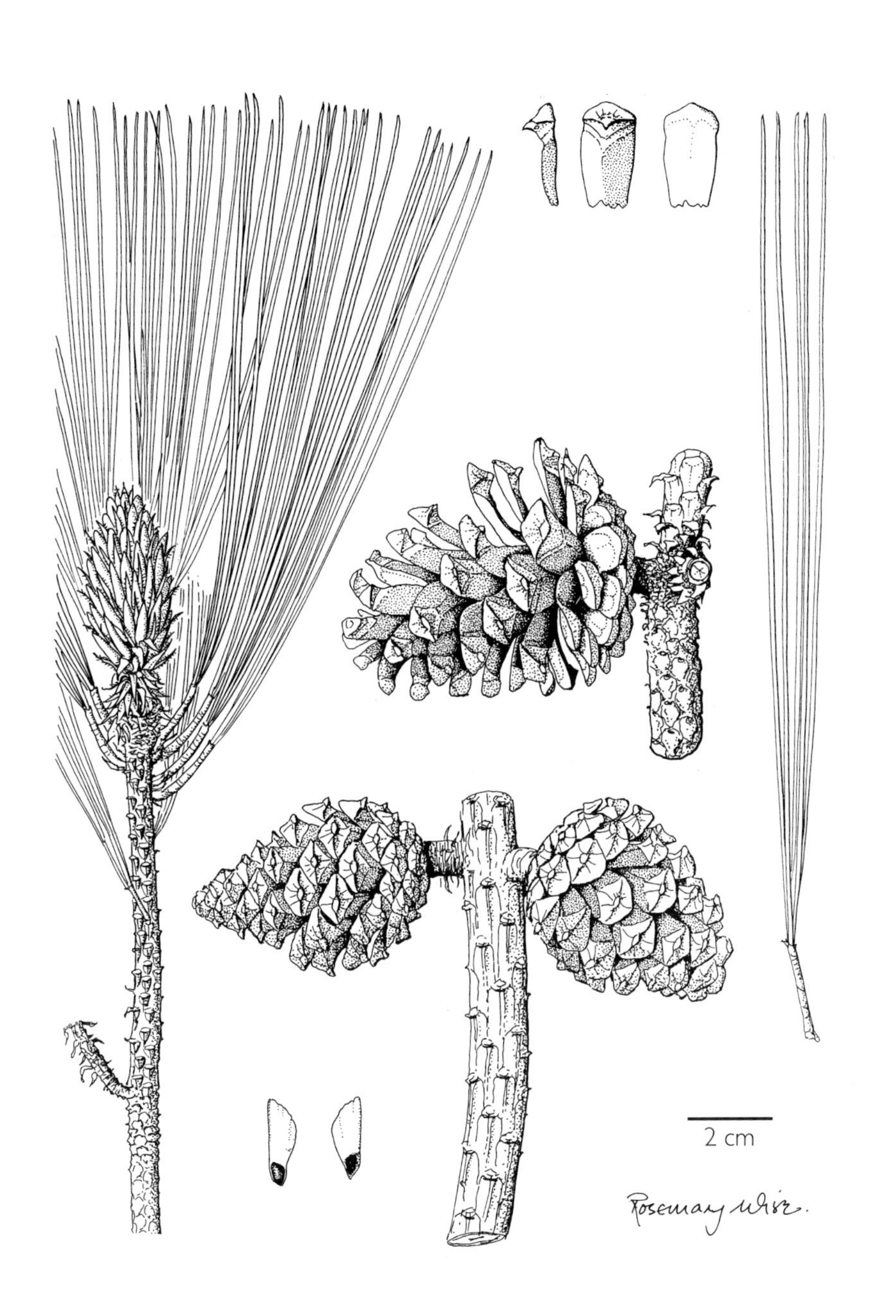
2 cm
Rosemary Wise.

26 Pinus lawsonii

Pinus lawsonii G. Gordon & Glendinning

Local names: Ocote, Pino chino, Pino ortiguillo

Habit, trunk: tree to 25–30 m tall, d.b.h. to 75 cm, trunk straight, sometimes tortuous.

Bark: thick on the trunk, scaly, with broad and deep longitudinal fissures, inner bark dark purplish-red, outer bark blackish-brown.

Foliage twigs: smooth, ridged, with small leaf bases (fascicle bases), orange-brown, often glaucous; needle fascicles spreading, persisting 2–3 years.

Needles: in fascicles of 3–4(–5), rarely 2, 12–20(–25) cm long, 1.0–1.2(–1.5) mm wide, straight, rigid, glaucous-green.

Cones: solitary or in pairs on short, stout peduncles, deciduous, ovoid-oblong, asymmetrical or symmetrical, 5–8(–9) × 4–6(–7) cm when open.

Cone scales: 70–100, gradually opening, thick woody, rigid, becoming recurved; apophysis slightly raised, transversely keeled, rhombic in outline, light brown, with an abruptly raised and curved umbo.

Seeds: 4–5 mm long, dark brown, with an articulate wing 12–16 × 5–6 mm.

Habitat: montane pine-oak and pine forests. This species is usually growing scattered with other pine species, on sites with sandy, shallow soil also with *Juniperus*.

Distribution: MEXICO: infrequent in S Mexico, in the states of Michoacán, México, Morelos, D.F., (one locality in) Veracruz, Guerrero and Oaxaca.

Altitude: 1300–2600 m.

Related or similar species: A number of species: *P. durangensis* (No. 25), *P. lawsonii* (No. 26), *P. pringlei* (No. 27) and *P. teocote* (No. 28) are probably closely related (Farjon & Styles, Flora Neotropica Monograph 75) and are similar in many characters. *Pinus lawsonii* is similar to *P. durangensis* in its raised umbos on the cone scales, but it has fewer needles per fascicle. The umbos of the other two species are flat.

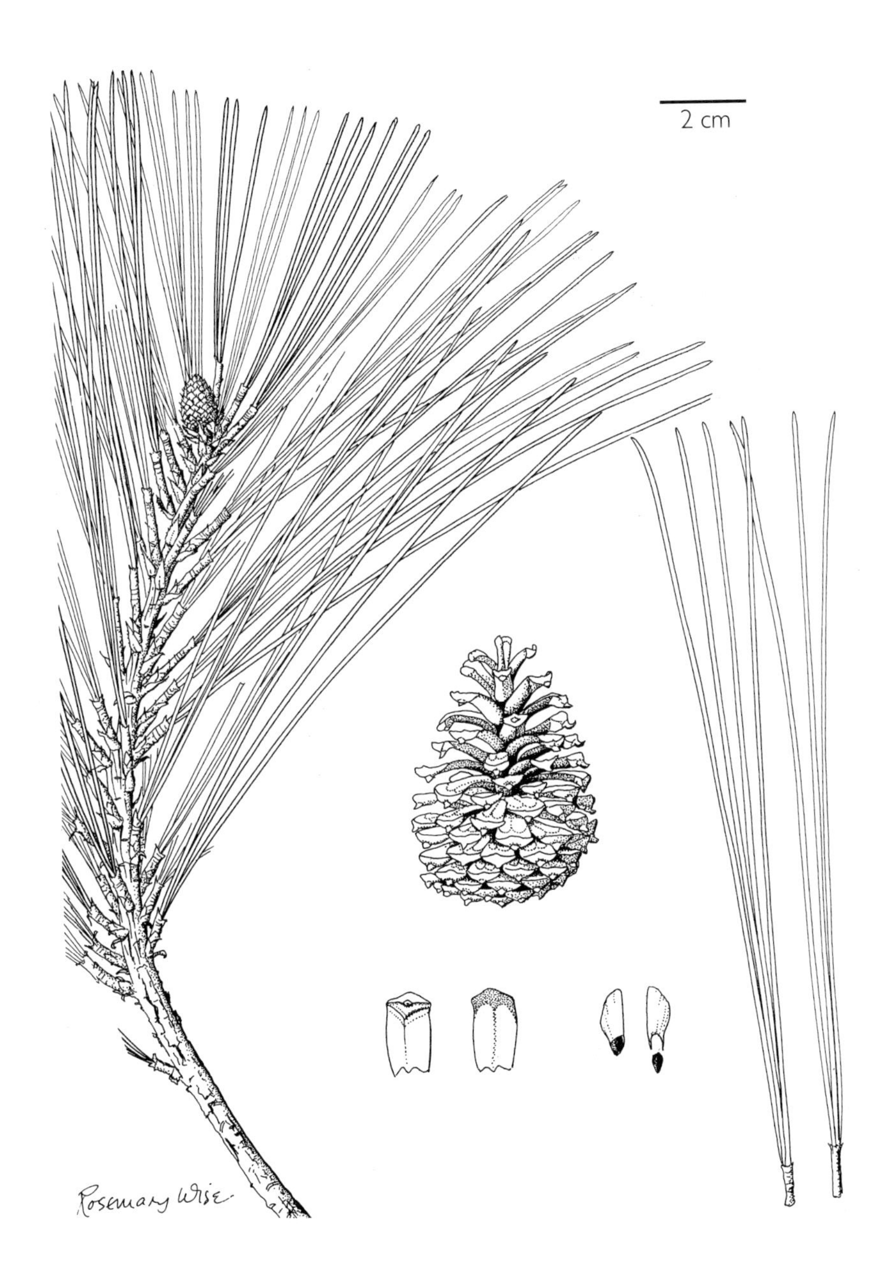
2 cm
Rosemary Wise

27 Pinus pringlei

Pinus pringlei G. R. Shaw

Local names: Ocote, Pino, Pino rojo

Habit, trunk: tree to 20–25 m tall, d.b.h. to 90–100 cm, trunk straight.

Bark: thick on the trunk, scaly, breaking into small plates, fissured, inner bark orange-red, outer bark greyish-brown, bark on branches exfoliating abundantly.

Foliage twigs: thick, smooth and ridged, new shoots glaucous, later reddish-brown, leaf bases (fascicle bases) soon eroding; needle fascicles spreading, persisting 2–3 years.

Needles: in fascicles of 3(–4), (15–)18–25(–30) cm long, 1.0–1.5(–1.7) mm wide, straight, rigid.

Cones: solitary or in whorls of 2–4 on short, stout peduncles, persistent, ovoid or slightly curved, 5–8(–10) × 3.5–6(–7) cm when open.

Cone scales: 70–100, slowly opening, those near the base often remaining closed, thick woody, rigid; apophysis flat to slightly raised on one side of the cone, light brown, with a flat or depressed umbo.

Seeds: 4–6 mm long, dark brown to grey-black, with an articulate, light brown wing 14–18 × 6–8 mm.

Habitat: montane pine and pine-oak forests. This species grows usually mixed with other pine species, on dry sites and in degraded forest often with *P. devoniana* and *P. lawsonii*.

Distribution: MEXICO: in southern Mexico in the states of Michoacán, México, Morelos, Guerrero and Oaxaca, perhaps in W Puebla.

Altitude: 1500–2600(–2800) m.

Notes: A "grass stage" has been reported for seedlings of this species.

Related or similar species: A related and similar species is *P. teocote* (No. 28), but it has more slender twigs and symmetrical, ovoid cones of which the scales open sooner and more completely. The needles of *P. teocote*, also predominantly in fascicles of 3, are shorter than the needles of *P. pringlei*. *Pinus durangensis* (No. 25) and *P. lawsonii* (No. 26) have raised (not flat) umbos on the cone scales and more needles in most fascicles.

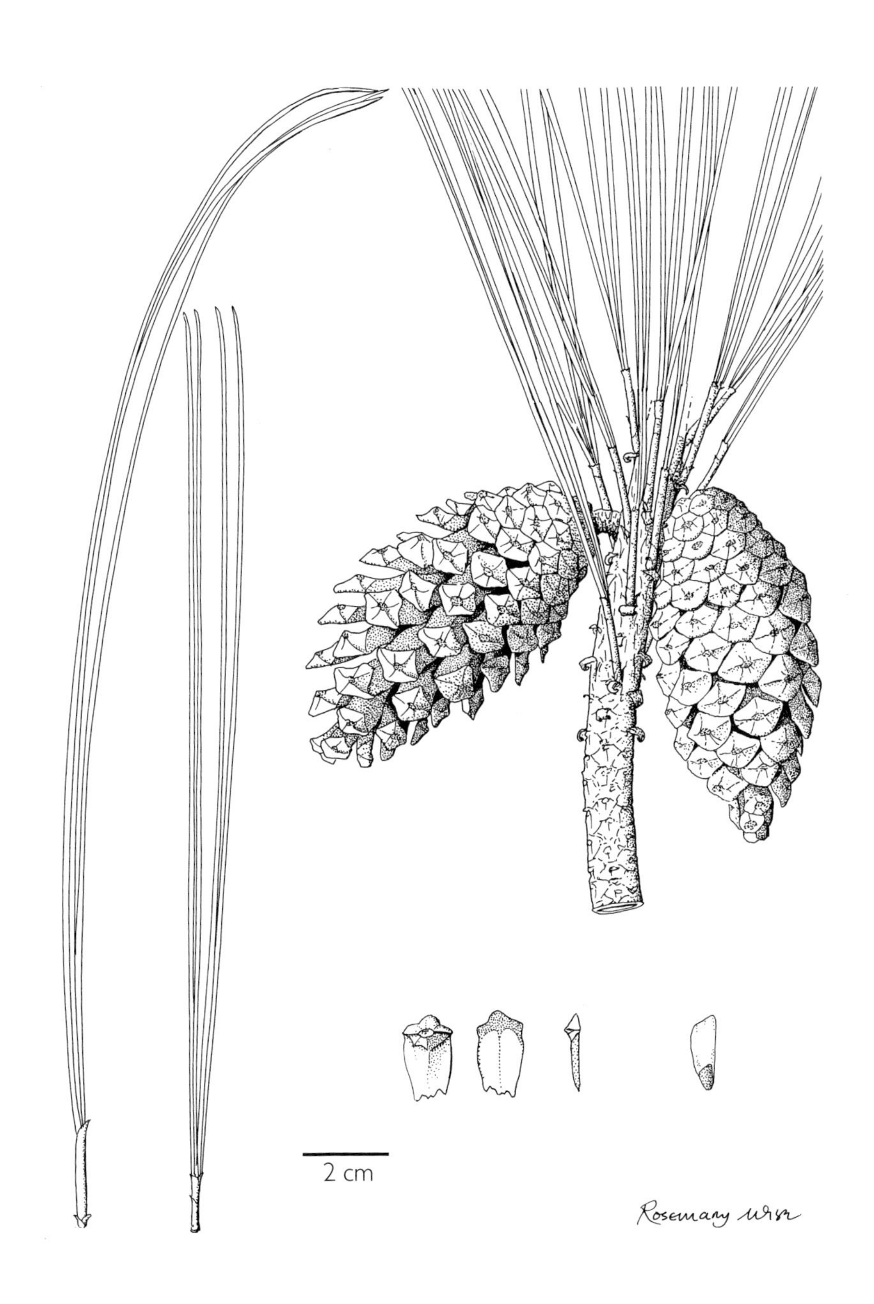
2 cm
Rosemary Wise

28 Pinus teocote

Pinus teocote Schlechtendal & Chamisso

Synonyms: *P. teocote* var. *macrocarpa* G. R. Shaw [= *P. teocote* forma *macrocarpa* (G. R. Shaw) Martínez]

Local names: Ocote, Pino chino, Pino colorado, Pino rosillo, Pino real

Habit, trunk: tree to 20–25 m tall, d.b.h. to 75 cm, trunk straight, sometimes forked.

Bark: thick on the trunk, scaly, with longitudinal plates and deep, wide fissures, inner bark reddish-brown, outer bark greyish-brown.

Foliage twigs: slender, with prominent leaf bases (fascicle bases), orange-brown; needle fascicles spreading, persisting 2–3 years.

Needles: in fascicles of 3(2–5), (7–)10–15(–18) cm long, 1–1.4 mm wide, straight or slightly curved towards the end of branchlets, rigid, dark green or light green; fascicle sheaths become much shorter with age.

Cones: in pairs, sometimes 1–3, on short, curved peduncles which fall with the cones, ovoid, slightly asymmetrical, (3–)4–6(–7) × 2.5–5 cm when open.

Cone scales: 60–100, opening soon, thick woody, rigid; apophysis flat to slightly raised, light brown, with a darker, flat or obtuse umbo.

Seeds: 3–5 mm long, dark grey-brown, with a lighter coloured, translucent, articulate wing 12–18 × 6–8 mm.

Habitat: relatively open and dry pine-oak and pine forest and woodland, often on dry, rocky ridges. This species also occurs in mixed broadleaved forest in the southern parts of its range, on areas with shallow and occasionally calcareous soil.

Distribution: MEXICO: widely distributed, most abundant in Central Mexico, rare and very scattered in the Sierra Madre Occidental. Reported from Guatemala but not confirmed by recent investigations.

Altitude: (1000–)1500–3000(–3300) m.

Related or similar species: *Pinus pringlei* (No. 27) has thicker foliage twigs (shoots) and longer needles; the cones are usually larger and more oblong when closed. *Pinus lawsonii* (No. 26) has cone scales with prominent, curved umbos; *P. patula* (No. 22) has very lax and slender needles and persistent cones. *Pinus durangensis* (No. 25) has usually 5 or more needles per fascicle and the cone scales have a prominent umbo.

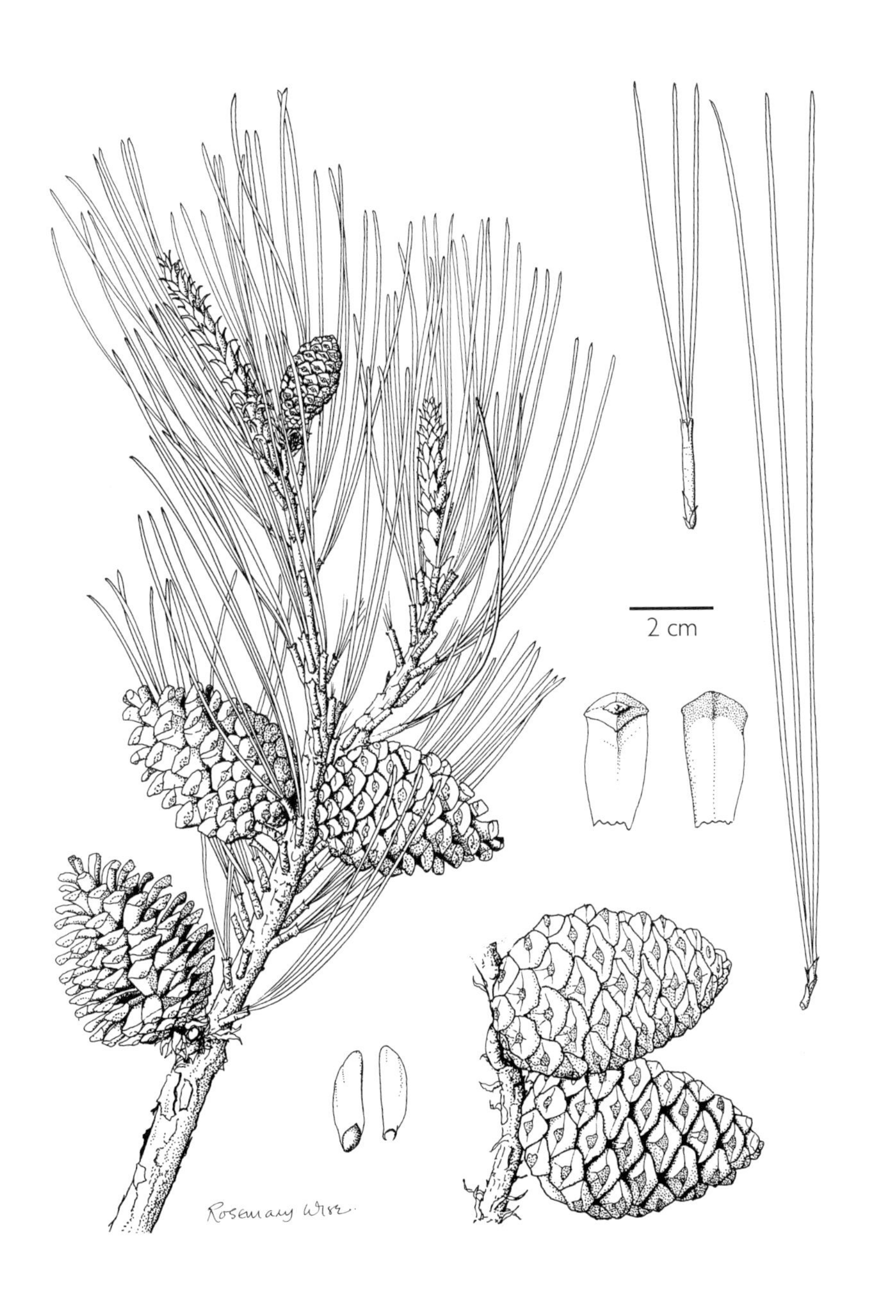
2 cm
Rosemary Wise

29 Pinus muricata

Pinus muricata D. Don var. *muricata*

Synonyms: *P. remorata* H. Mason

Local names: Pino

Habit, trunk: shrub or small tree to 4–10(–15?) m tall, d.b.h. 20–50 cm, trunk erect, often curved, branching low.

Bark: rough and scaly, on larger trunks with deep longitudinal fissures, dark brown to grey.

Foliage twigs: (main) shoots multinodal, with large, persistent leaf bases (fascicle bases); needle fascicles spreading, persisting 2–3 years.

Needles: in fascicles of 2, (7–)10–14(–16) cm long, 1.3–2.0 mm wide, straight or slightly curved, very rigid.

Cones: solitary or in whorls of 2–5 on (very) short peduncles, persistent, narrowly ovoid or asymmetrical with long, spiny apophyses, 5–7(–8) × 4–5(–6) cm when (half) open.

Cone scales: 70–100, remaining closed or opening very slowly, thick woody, rigid; apophysis very variable, from slightly raised to extremely elongated on one side of the cone, with an obtuse or acute, curved umbo armed with a sharp prickle.

Seeds: 5–6 × 3–4.5 mm, grey to black, with an articulate, lighter coloured wing 14–18 × 5–8 mm.

Habitat: the coastal chaparral zone, on north-facing slopes or steep escarpments within the fog-belt, mostly with volcanic rock or shallow rocky soil. This species forms open groves, sometimes mixed with *Cupressus* or *Juniperus*, and is adapted to frequent brush fires.

Distribution: MEXICO: Baja California Norte, in two localities on the mainland near the coast, W and SW of San Vicente; the largest groves are on and near Cerro Colorado on the north side of the Rio San Isidro. **Not** on Isla Cedros: the pine there is *P. radiata* var. *binata* (No. 30), but also in several locations on the coast of Alta California (U.S.A.).

Altitude: 1–100 m (in Mexico).

Related species, subspecies or varieties: This is one of the three Californian "closed-cone" pines, the two other species are *P. attenuata* (No. 31) and *P. radiata* (No. 30). *Pinus attenuata* has needles in fascicles of 3 (rarely 2) and larger, oblong cones; *P. radiata* var. *binata* occurs only on Isla Cedros and Isla Guadalupe, not on the mainland of Baja California.

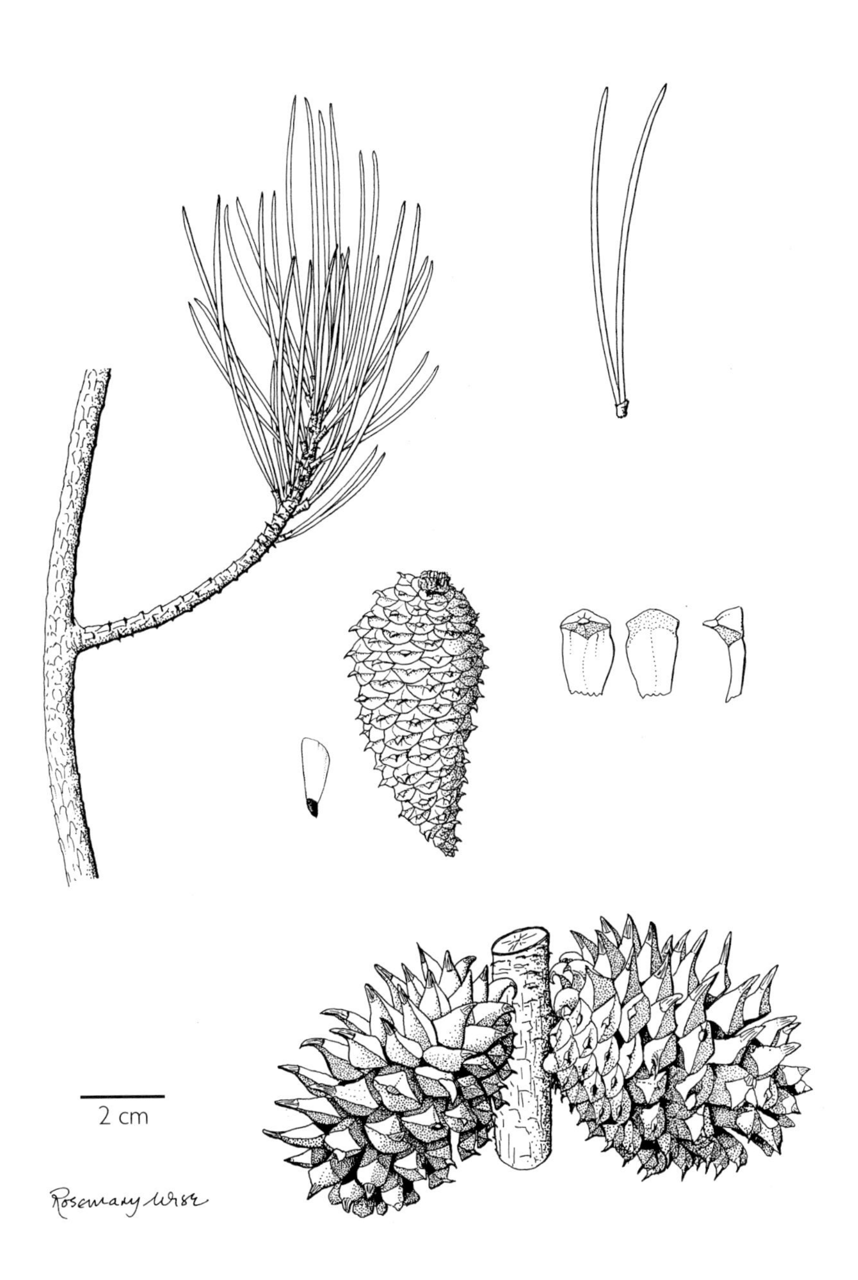
2 cm
Rosemary Wise

30 Pinus radiata var. binata

Pinus radiata D. Don var. *binata* (Engelmann) J. G. Lemmon

Synonyms: *P. radiata* forma *guadalupensis* J. T. Howell; *P. muricata* var. *cedrosensis* J. T. Howell; *P. radiata* var. *cedrosensis* (J. T. Howell) Silba

Local names: Pino

Habit, trunk: small to medium sized tree to 20–25(–33) m tall, d.b.h. to 200–220 cm, trunk massive, erect, often branching low.

Bark: very thick on the trunk, scaly, deeply fissured between large plates, inner bark brown, outer bark blackish-grey.

Foliage twigs: (main) shoots multinodal, thick, with prominent leaf bases (fascicle bases); needle fascicles spreading, persisting 3 years.

Needles: in fascicles of 2 (2–3 on leading shoots), 8–15 cm long, 1.1–1.6 mm wide, straight or curved to twisted, rigid.

Cones: solitary or in whorls of 2–5 on very short peduncles, persistent, ovoid, (5–)6.5–9(–13) × (4–)6–7(–9) cm when open.

Cone scales: 130–160, opening very slowly, thick woody, rigid; apophysis slightly raised, transversely keeled, or more thickened near the base on one side of the cone, light brown, with a flat, grey umbo.

Seeds: (5–)6–8 × 4–5 mm, grey-brown with dark spots; wing articulate, yellowish-brown, translucent, 14–20 × 7–9 mm.

Habitat: rocky slopes or moist canyons on the windward side of two islands in the Pacific Ocean, in an extremely oceanic climate with daily fog and intermittent rain.

Distribution: MEXICO: Baja California Norte, on Isla Cedros and Isla Guadalupe, situated respectively near the mainland coast and 250 km off the coast. It is the only pine on these islands.

Altitude: (200–)300–650 m (Isla Cedros); 300–1160 m (Isla Guadalupe).

Notes: Martínez (1948) and other authors have referred to the pines of Isla Cedros as (a variety of) *P. muricata* (No. 29), but more detailed observations have shown that this species only occurs on the mainland.

Related species, subspecies or varieties: In Alta California (U.S.A.) occurs *P. radiata* var. *radiata*, the Monterey pine; var. *binata* differs only slightly from it, with on average smaller and more symmetrical cones and only occasional fascicles of 3 needles on leading shoots.

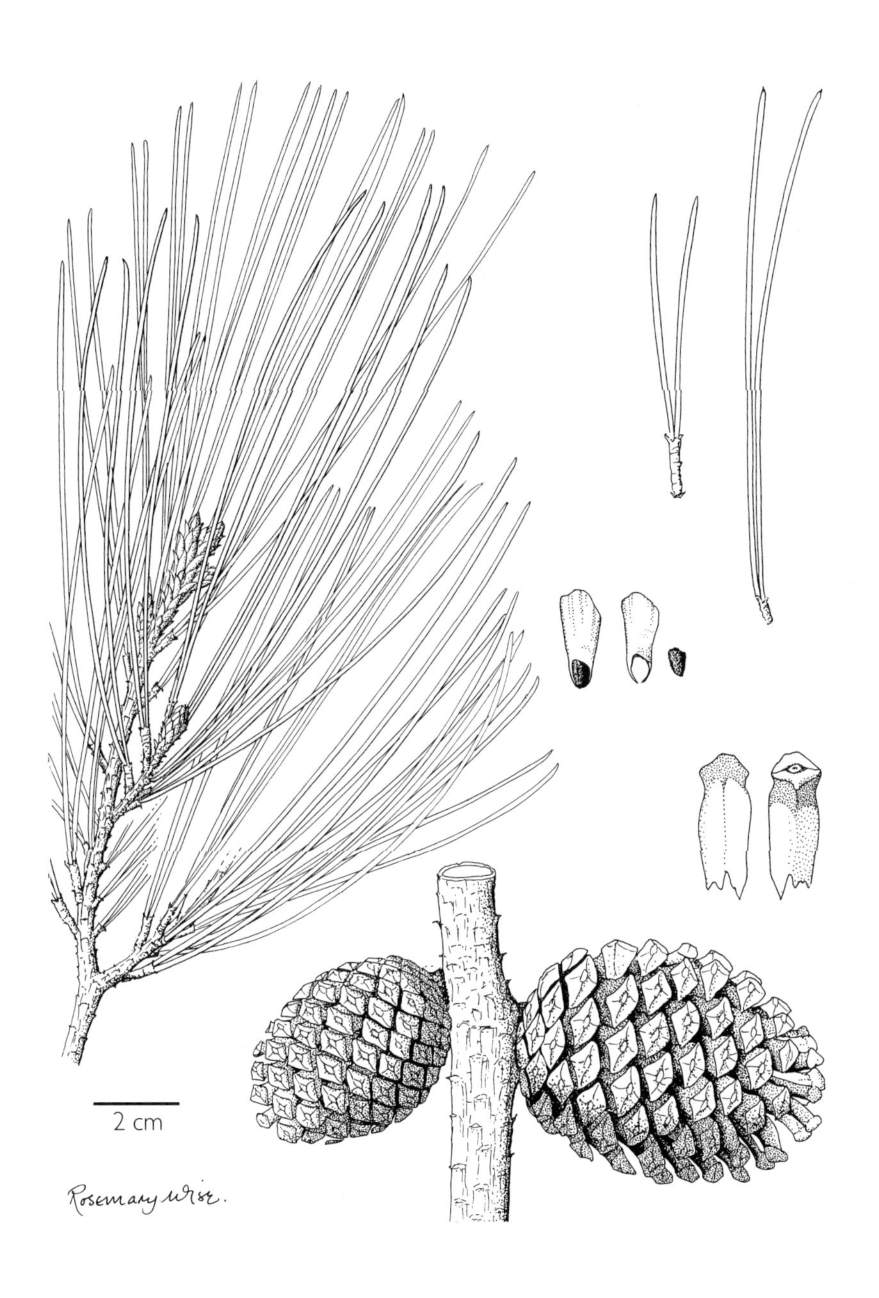
2 cm
Rosemary Wise.

31 Pinus attenuata

Pinus attenuata J. G. Lemmon

Local names: Chichonuda, Pino de piña

Habit, trunk: tree to 15–20(–25) m tall, d.b.h. to 40–50 cm, trunk straight or curved, often branched low.

Bark: relatively thin, scaly, breaking into small, rectangular plates, greyish-brown to grey.

Foliage twigs: (main) shoots often multinodal, with persistent leaf bases (fascicle bases); needle fascicles spreading, persisting 2–3 years.

Needles: in fascicles of 3, rarely 2, (8–)10–12(–14) cm long, 1.0–1.5 mm wide, straight, rigid.

Cones: solitary or more often in whorls of 2–5, nearly sessile, very persistent on main stem and branches, reflexed, ovoid-oblong to oblong, 8–15 × 3.5–6 cm when closed (up to 8 cm wide when open).

Cone scales: 150–180, remaining closed for many years; apophysis from slightly raised on the side of the branch to conical and curved near the base on the upper side, yellowish-brown, with an obtuse, raised or flat, darker umbo.

Seeds: 5–7 × 3.5–4.5 mm, slightly acute, blackish-grey, with an articulate wing 12–18 × 5–7 mm.

Habitat: dry rocky slopes in the chaparral zone of the lower slopes of coastal mountains, where brush fires are frequent.

Distribution: MEXICO: Baja California Norte, a few scattered stands in the coastal mountains around Ensenada; it is more common and widespread in the coastal regions of Alta California and Oregon (U.S.A.).

Altitude: 250–600 m (in Mexico).

Notes: The populations near Ensenada represent the southernmost distribution of this species.

Related species: This species is one of the (Californian) "closed-cone" pines; besides *P. muricata* (No. 29) and *P. radiata* (No. 30) of Alta California and parts of Baja California, *P. greggii* (No. 32) belongs to this group (Farjon & Styles, Flora Neotropica Monograph 75). The oblong, tapering and strictly closed cones with "knobby" apophyses on the upper side towards the base, combined with needles in fascicles of 3, distinguish *P. attenuata*.

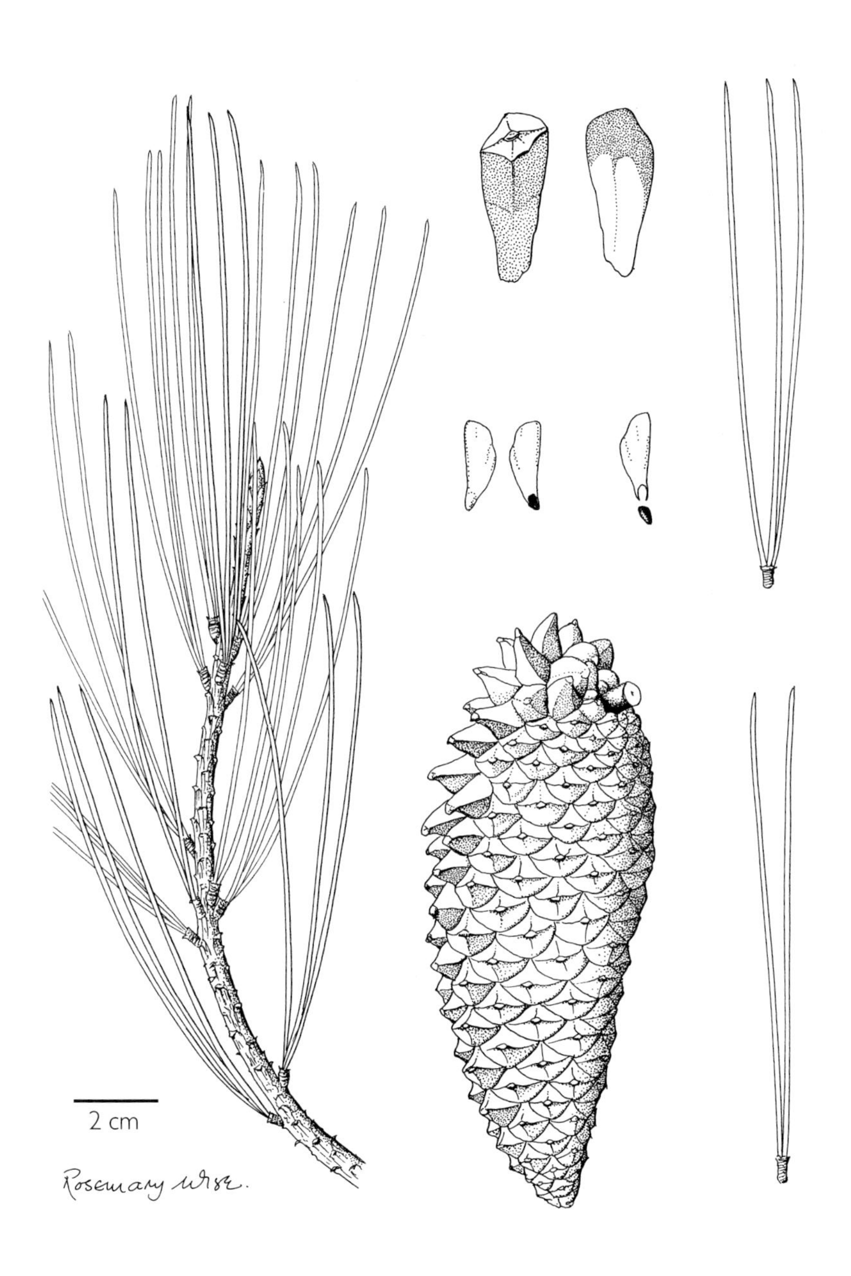
2 cm
Rosemary Wise.

32 Pinus greggii

Pinus greggii Parlatore

Local names: Pino chino, Pino prieto

Habit, trunk: tree to 20–25 m tall, d.b.h. to 70–80 cm, trunk straight.

Bark: thick on the trunk, scaly, with elongated plates and deep longitudinal fissures, dark brown, to greyish-brown outside.

Foliage twigs: (main) shoots often multinodal, smooth, ridged between leaf bases (fascicle bases), yellowish- to grey-brown; needle fascicles spreading forward, persisting up to 4 years.

Needles: in fascicles of 3, (7–)9–13(–15) cm long, 1–1.2 mm wide, straight and rigid.

Cones: appearing on very young trees, in whorls of (1–)3–8 or more, nearly sessile, persistent, narrowly ovoid to oblong when closed, with an oblique base, (6–)8–13(–15) × (4–)5–7 cm when open (width 3.5–5 cm when closed); remaining closed 4–8 years after maturing on the tree.

Cone scales: 80–120, remaining closed several to many years; apophysis flat or slightly raised, yellowish-brown, with a depressed or flat, grey umbo.

Seeds: 5–8 × 3–4 mm, grey to blackish-brown, with an articulate wing 15–20 × 6–8 mm.

Habitat: various types of montane mixed broadleaved forest or pine and pine-oak forest, on acidic soils or slightly alkaline soils in the north of the range.

Distribution: MEXICO: SE Coahuila, S Nuevo León, SE San Luis Potosí, Querétaro, Hidalgo and N Puebla; this species is nowhere abundant.

Altitude: 1300–2600 m, in the north 2300–2700 m.

Notes: This relatively rare species is apparently closely related to the Californian "closed-cone" pines *P. attenuata* (No. 31), *P. muricata* (No. 29) and *P. radiata* (No. 30) (see Farjon & Styles, Flora Neotropica Monograph 75).

Similar species: The Californian "closed-cone" pines extend into Baja California and two Pacific Islands; a species that is more or less similar (but not closely related) and occurs on the Mexican mainland is *P. patula* (No. 22). *P. patula* differs in its longer, slender, lax and drooping to pendulous needles in fascicles of 3–4(–5) and its smaller, less serotinous cones.

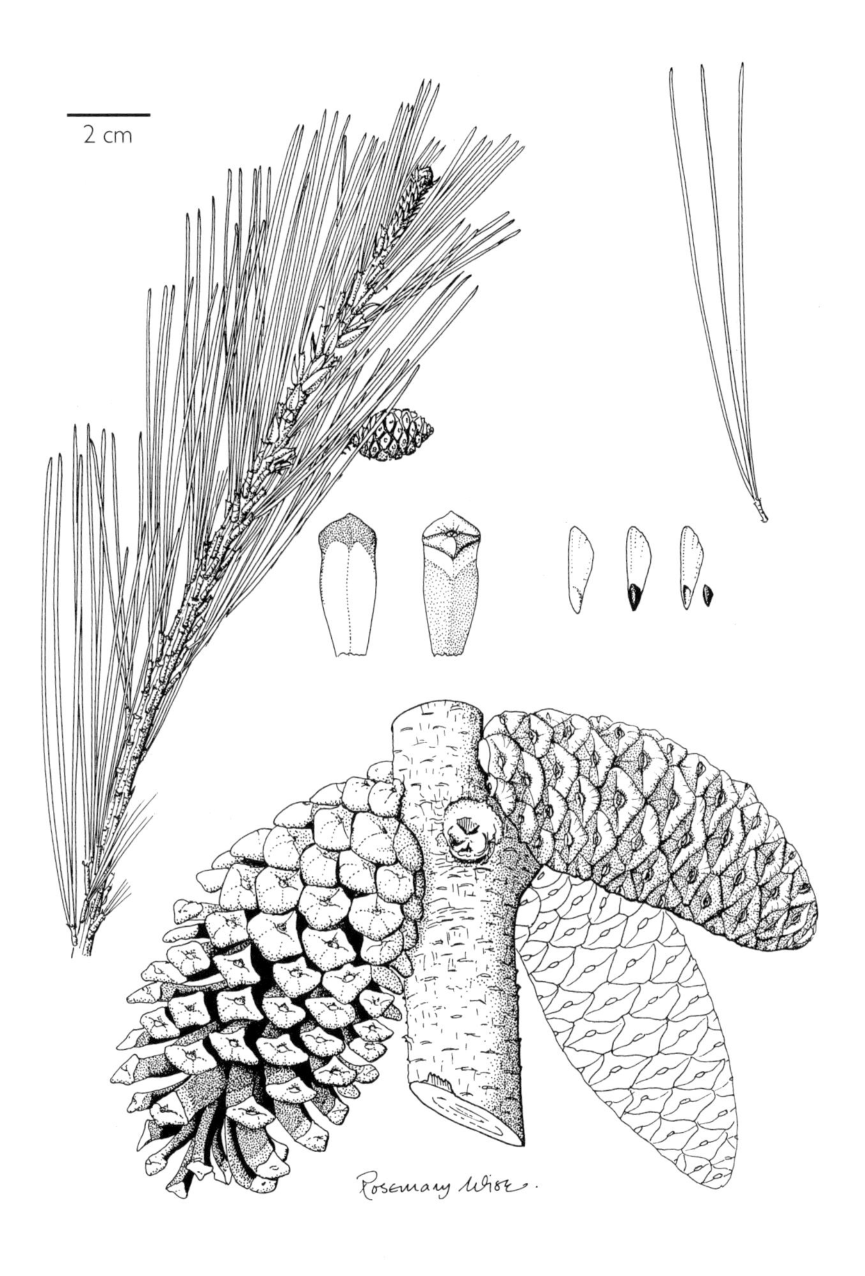
2 cm
Rosemary Wise.

33 Pinus coulteri

Pinus coulteri D. Don

Local names: Pino

Habit, trunk: tree to 15–25 m tall, d.b.h. to 100 cm, trunk straight or curved at base, main branches very long.

Bark: thick on the trunk, scaly, with irregular, longitudinal fissures and large plates, dark brown with black fissures.

Foliage twigs: (main) shoots multinodal, thick, rough, with prominent leaf bases (fascicle bases), light orange-brown, often glaucous; needle fascicles spreading, persisting 3–4 years.

Needles: in fascicles of 3, 15–25(30) cm long, 1.9–2.2 mm wide, straight or curved, very rigid, often resinous.

Cones: solitary or in pairs, sometimes in whorls of 3–4(–5) on stems of young trees, on thick, short peduncles, ovoid, massive, 20–35 × 15–20 cm when open, extremely resinous.

Cone scales: 180–220, thick woody, rigid, opening slowly; apophysis very strongly developed, elongated and curved, with a long, hooked and sharp umbo, light yellowish-brown.

Seeds: 10–18 × 7–10 mm, dark brown to blackish, with an articulate, brown wing 18–30 × 12–16 mm thickened at the base.

Habitat: lower limits of pine forest, extending into chaparral, or rocky areas among granite boulders; this pine forms open stands, often with *Quercus chrysolepis*.

Distribution: MEXICO: Baja California Norte, scattered and rare in the Sierra Juárez and Sierra San Pedro Martír; more widespread and abundant in the Coast Ranges of Alta California (U.S.A.).

Altitude: 1200–1800 m (Sierra Juárez), 1900–2150 m (Sierra San Pedro Martír).

Notes: Natural hybrids may occur between *P. coulteri* and *P. jeffreyi* where the two species grow together; such hybrids have been reported from Alta California but not yet from Baja California.

Related species: Although hybridization seems to point at *P. jeffreyi* (No. 11) as a related species, phylogenetic analysis indicates a close relationship with the Californian "closed-cone" pines *P. attenuata* (No. 31), *P. muricata* (No. 29) and *P. radiata* (No. 30) instead (Farjon & Styles, Flora Neotropica Monograph 75). In Alta California occur two other close relatives: *P. sabiniana* and (rare) *P. torreyana*. In Mexico, no other pine has large cones with such heavy, hooked apophyses.

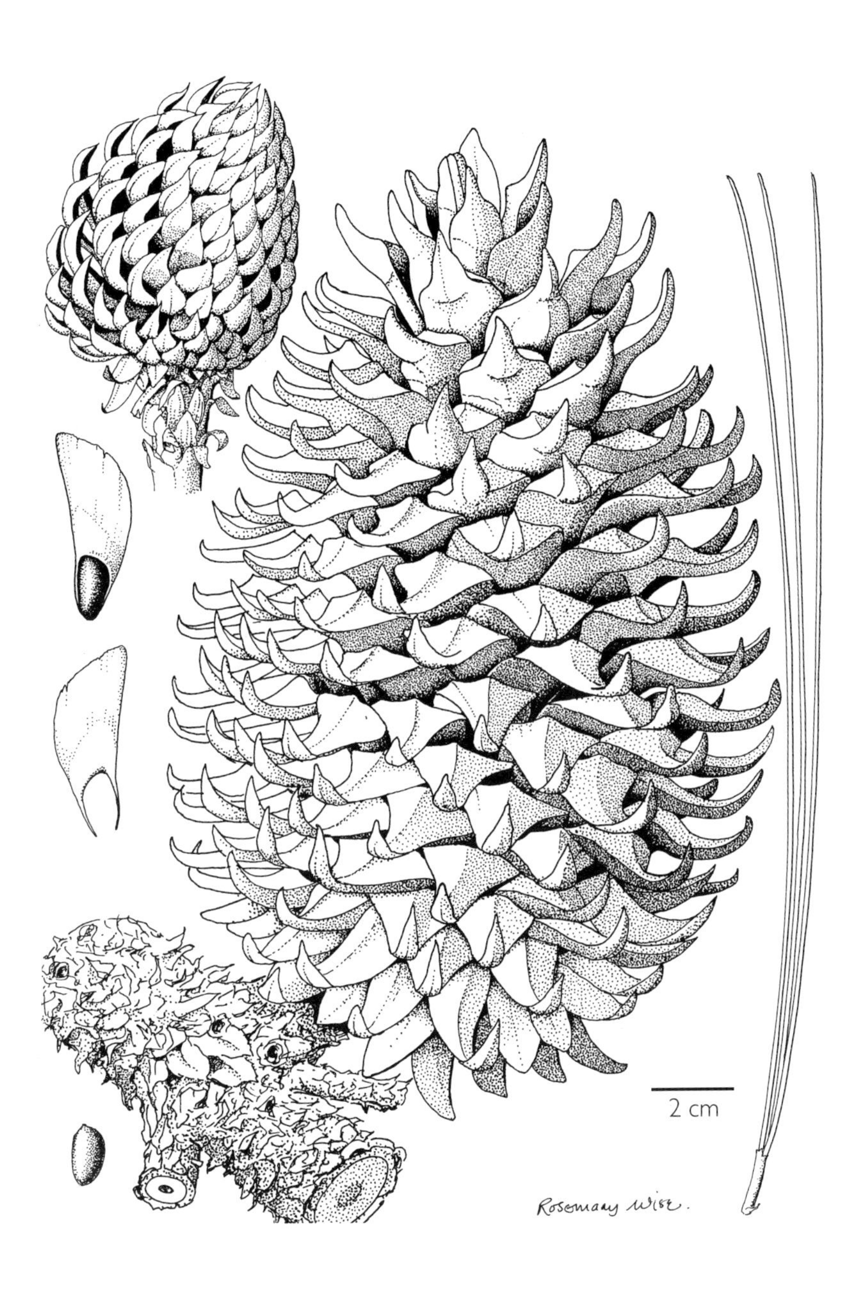
2 cm
Rosemary Wise.

34 Pinus ayacahuite

Pinus ayacahuite Schlechtendal var. *ayacahuite*

Local names: Acalocote, Ayacahuite, Ocote gretado, Pinabete, Pino gretado, Pino tabla

Habit, trunk: tree to 40–45 m tall, d.b.h. to 150–200 cm, trunk straight, with spreading branches along $^{2}/_{3}$–$^{3}/_{4}$ of length.

Bark: not very thick on the trunk, scaly, divided into small, rectangular scales, grey-brown to grey; on young trees thin and smooth.

Foliage twigs: slender, young shoots may be slightly pubescent, leaf bases (fascicle bases) prominent, short decurrent; needle fascicles spreading but lax, persisting 2–3 years.

Needles: in fascicles of 5, (8–)10–15(–18) cm long, 0.7–1.0 mm wide, straight or slightly curved, lax, margins (weakly) serrulate, stomata only on two inner sides; scales of fascicle sheaths soon falling.

Cones: solitary or in whorls of 2–4 on short peduncles, pendulous, deciduous, cylindrical, curved, (10–)15–40 × 7–15 cm when open, dull brown.

Cone scales: 100–150, opening soon and widely, thin, flexible; apophysis irregular, often recurved or reflexed, with a terminal, obtuse umbo, very resinous.

Seeds: 8–10 × 6–8 mm, with an adnate wing 20–35 × 8–12 mm.

Habitat: mixed montane conifer forest on mesic sites. This pine is often an emergent tree, or it occurs in small groves, best developed on loamy, well drained soils.

Distribution: MEXICO: (including the variety below) Guanajuato, Quéretaro, Hidalgo, Puebla, Veracruz, Tlaxcala, Mexico, Morelos, Michoacán, Guerrero, Oaxaca, Chiapas; GUATEMALA, throughout the highlands; HONDURAS: on highest mountains; EL SALVADOR: only in Chalatenango.

Altitude: (1500–)1900–3200(–3600) m.

Notes: This species is a very important timber tree in Mexico and Guatemala, due to its size, quality of shape and wood.

Related species, subspecies or varieties: In parts of the range in Central Mexico, *P. ayacahuite* var. *veitchii* (Roezl) G. R. Shaw is distinguished by its often very large cones (up to 50 cm long) with extremely elongated and reflexed apophyses and seed wings up to 2x the length of the seeds. *Pinus strobiformis* (No. 37), sometimes regarded as a variety of *P. ayacahuite*, is here treated as a distinct species (see Farjon & Styles, Flora Neotropica Monograph 75).

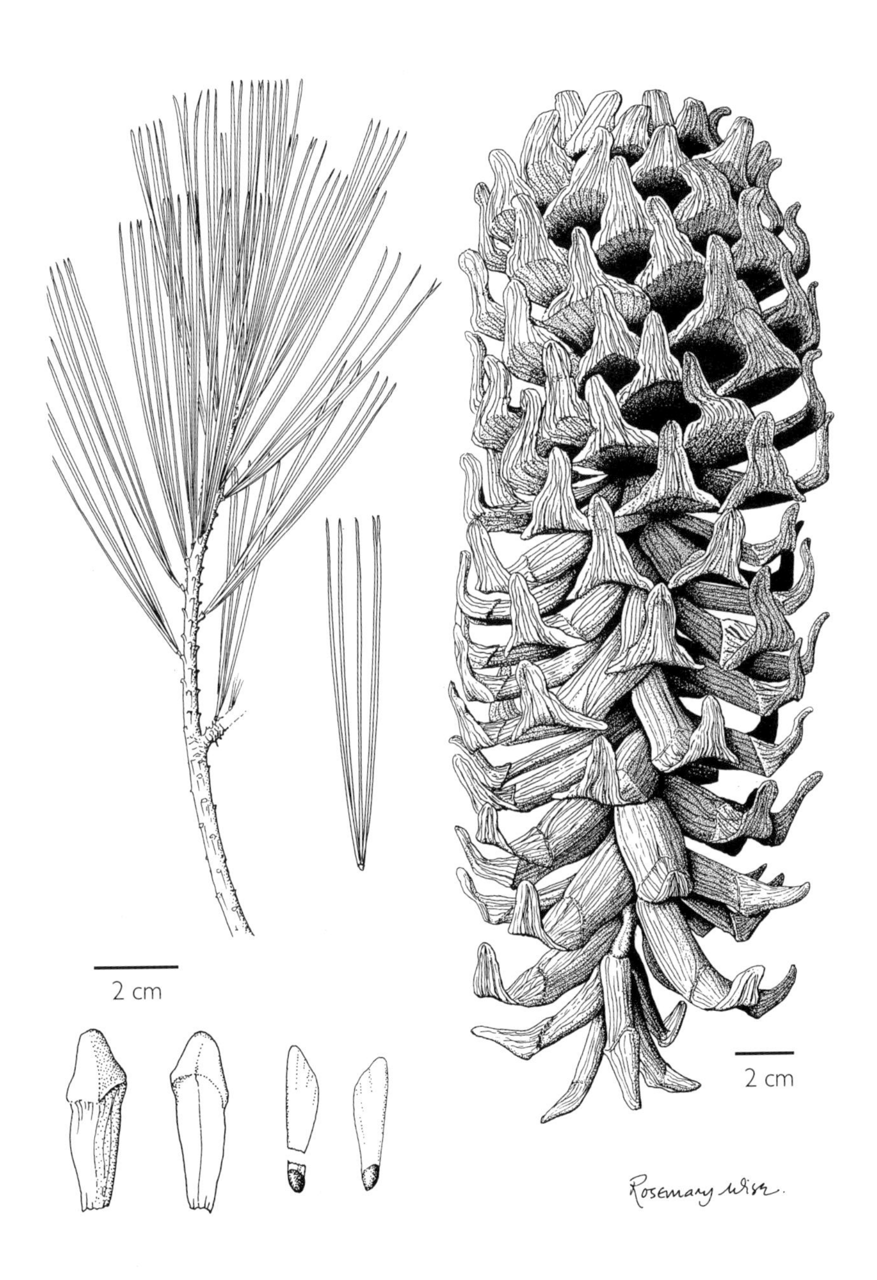
2 cm
2 cm
Rosemary Wise

35 Pinus lambertiana

Pinus lambertiana D. Douglas

Local names: Ocote, Pino de azúcar

Habit, trunk: tree to 30–40 m tall, d.b.h. 80–120 cm, trunk straight, branches long and spreading.

Bark: thick on the trunk, on the lower part with irregular, large plates and deep fissures, grey-brown; on young trees thin and smooth.

Foliage twigs: slender, flexible, young shoots slightly pubescent, with small leaf bases (fascicle bases), orange-brown, soon light grey; needle fascicles spreading but lax, persisting 2–4 years.

Needles: in fascicles of 5, (3.5–)4–8(–10) cm long, 0.8–1.5 mm wide, straight, lax, very weakly serrulate, with stomata on all three sides; scales of fascicle sheaths soon falling.

Cones: near the ends of main branches, solitary or in whorls of 2–4, pendulous, deciduous, cylindrical, nearly straight, 25–45 × 8–14 cm when open, light brown.

Cone scales: 110–130, opening soon and wide, up to 4–5 mm thick, rigid; apophysis obtuse triangular, 5–8 mm thick at base, not recurved, with a terminal, obtuse umbo, very resinous.

Seeds: (10–)12–15(–18) × 6–10 mm, dark brown, with an adnate, light brown wing 20–30 × 12–15 mm.

Habitat: mixed montane conifer forest, on sites with deepest soil. This species grows often along intermittent streams, associated with *Abies concolor*, *Pinus contorta* var. *murrayana* and *P. jeffreyi*.

Distribution: MEXICO: Baja California Norte, Sierra San Pedro Martír; widespread in Alta California and Oregon (U.S.A.).

Altitude: 2200–2800 m (in Mexico).

Notes: In Mexico, this species does not attain the great size of trees found in Alta California and Oregon; references to sizes of 60 m or more in Martínez (1948) and Perry (1991) refer to the species outside Mexico. Likewise, the largest cones (56 cm long) have only been found in the U.S.A.

Related species: This species is the only representative in Baja California of the "White pines" (*Pinus* subsection *Strobus*), the other Mexican species (*P. ayacahuite* No. 34, *P. flexilis* var. *reflexa* No. 36, *P. strobiformis* No. 37 and *P. strobus* var. *chiapensis* No. 38) all occur on the mainland.

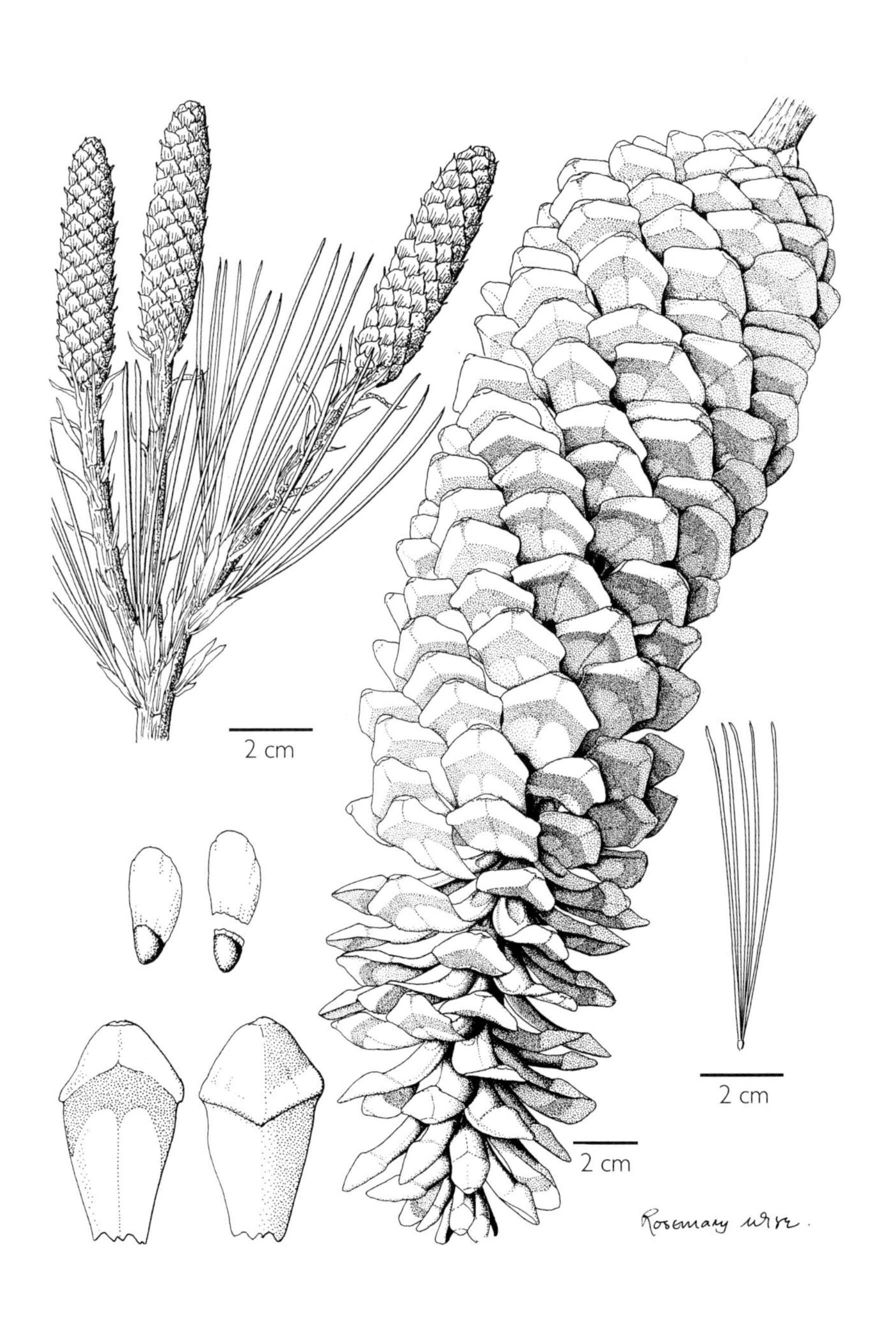
2 cm
2 cm
2 cm
Rosemary Wise

36 Pinus flexilis var. reflexa

Pinus flexilis E. James var. *reflexa* Engelmann

Synonyms: *P. reflexa* (Engelmann) Engelmann

Local names: Pinabete, Pino nayar?

Habit, trunk: tree to 10–15(–20) m tall, d.b.h. to 100–150 cm, with 1-several stems, branching very low, straight or contorted, branches curved upwards.

Bark: thin, scaly, with small scales and plates, brown to grey.

Foliage twigs: slender, flexible, young shoots slightly pubescent, grey-green, soon whitish-grey and smooth, with resinous winter buds; needle fascicles dense, spreading forward, persisting (3–)5–6 years.

Needles: in fascicles of 5, (5–)6–9 cm long, 0.8–1.2 mm wide, straight or twisted, lax, very weakly serrulate, with stomata mainly or entirely on the two inner sides; scales of fascicle sheaths soon falling.

Cones: solitary or in whorls of 2–4, on short peduncles, more or less pendulous, deciduous, cylindrical, straight or slightly curved, 10–15 × 4–6 cm when open, light brown.

Cone scales: 80–110, opening soon but not wide; apophysis thick, triangular or rhombic, obtuse, (slightly) recurved, with a terminal, obtuse umbo, very resinous.

Seeds: 10–15 × 8–10 mm, dark brown, with a rudimentary or very small, adnate wing up to 5 mm long; wing often absent when the seed is free from the scale.

Habitat: high montane to subalpine open woodland or exposed mountain ridges above the forest belt, with rocky soil and exposed to high winds and snowstorms.

Distribution: MEXICO: on a few mountain summits in Chihuahua, Coahuila and S Nuevo León; possibly more common.

Altitude: not recorded, but likely above 2500 m.

Related species, subspecies or varieties: *Pinus flexilis* and *P. strobiformis* (No. 37) are two closely related species occurring in the Cordillera of North America, from Canada into Mexico. Variations in especially the cones are greatest in the SW United States and North Mexico, where intermediate forms are sometimes encountered. *Pinus flexilis* var. *reflexa* is essentially similar to the subalpine *P. flexilis* var. *flexilis*, but its cone scales are slightly reflexed and it often has a small, ineffective seedwing. In the U.S.A. *P. flexilis* var. *reflexa* is commonly included in *P. strobiformis*.

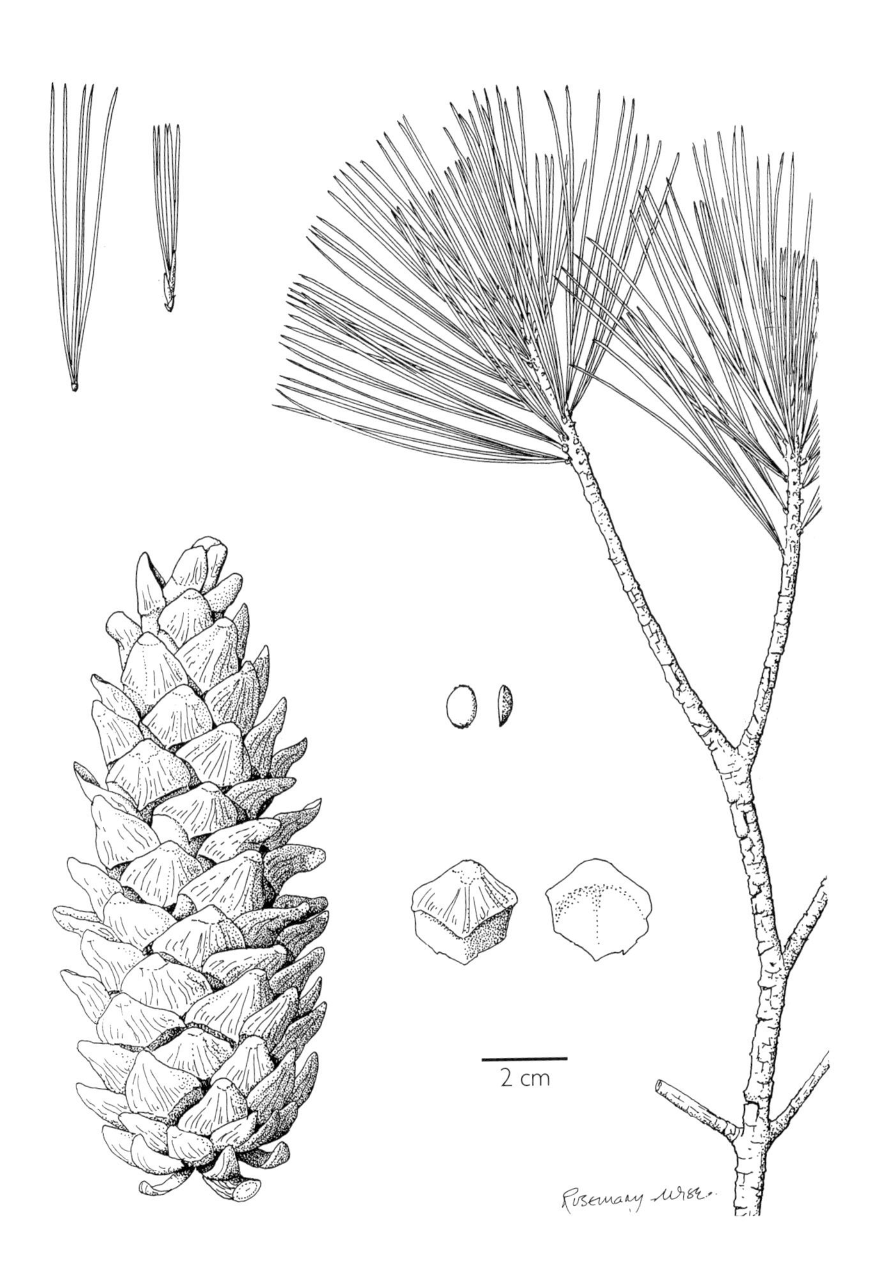
2 cm

37 Pinus strobiformis

Pinus strobiformis Engelmann

Synonyms: *Pinus ayacahuite* var. *brachyptera* Shaw; *P. ayacahuite* var. *novogaliciana* Carvajal

Local names: Acahuite, Pinabete, Pino blanco, Pino huiyoco, Pino nayar

Habit, trunk: tree to 25–30 m tall, d.b.h. to 80–100 cm, trunk straight.

Bark: thick on the trunk, scaly, breaking into small, irregular plates and fissures, dark brown, weathering grey; on young trees thin, smooth, grey.

Foliage twigs: slender, sometimes slightly pubescent, with prominent but small leaf bases (fascicle bases), yellowish-green to pale reddish-brown, soon grey; needle fascicles spreading but lax, persisting 3–5 years.

Needles: in fascicles of 5, (5–)7–11(–12) cm long, (0.6–)0.8–1.1(1.2) mm wide, weakly serrulate (sometimes more densely), straight or slightly twisted, lax; stomata only on the two inner sides; scales of fascicle sheaths soon falling.

Cones: solitary or in whorls of 2–4 on ca. 20 mm long peduncles, pendulous, deciduous, ovoid-oblong to cylindrical, long cones slightly curved, 12–30(–60) × 7–11 cm when open, light brown.

Cone scales: 70–120, opening soon and spreading wide, with deep seed cavities; apophysis triangular, thick at base, often elongated and recurved, yellowish-brown with a terminal, obtuse umbo, usually very resinous.

Seeds: 12–18 × 8–11 mm, brown, commonly only one of the seeds on a scale develops fully; wing adnate, rudimentary, up to $\frac{1}{2}$ as long as the seed.

Habitat: montane pine and pine-oak forest, mesic sites on relatively deep soil, also N-facing slopes and at high elevations conifer forest with *Abies* and/or *Pseudotsuga*.

Distribution: MEXICO: mainly in the Sierra Madre Occidental and Sierra Madre Oriental, in Sonora, Chihuahua, Coahuila, Nuevo León, Sinaloa, Durango, Jalisco and very locally in Zacatecas and San Luis Potosí; also in Arizona, New Mexico and W Texas (U.S.A.).

Altitude: 1900–3500 m.

Related species: Some botanists regard the more southern populations of this species as a variety of *P. ayacahuite* (No. 34), but a more detailed study by Perez de la Rosa has recently established that these, too belong with *P. strobiformis*. The cones are very variable in size and shape.

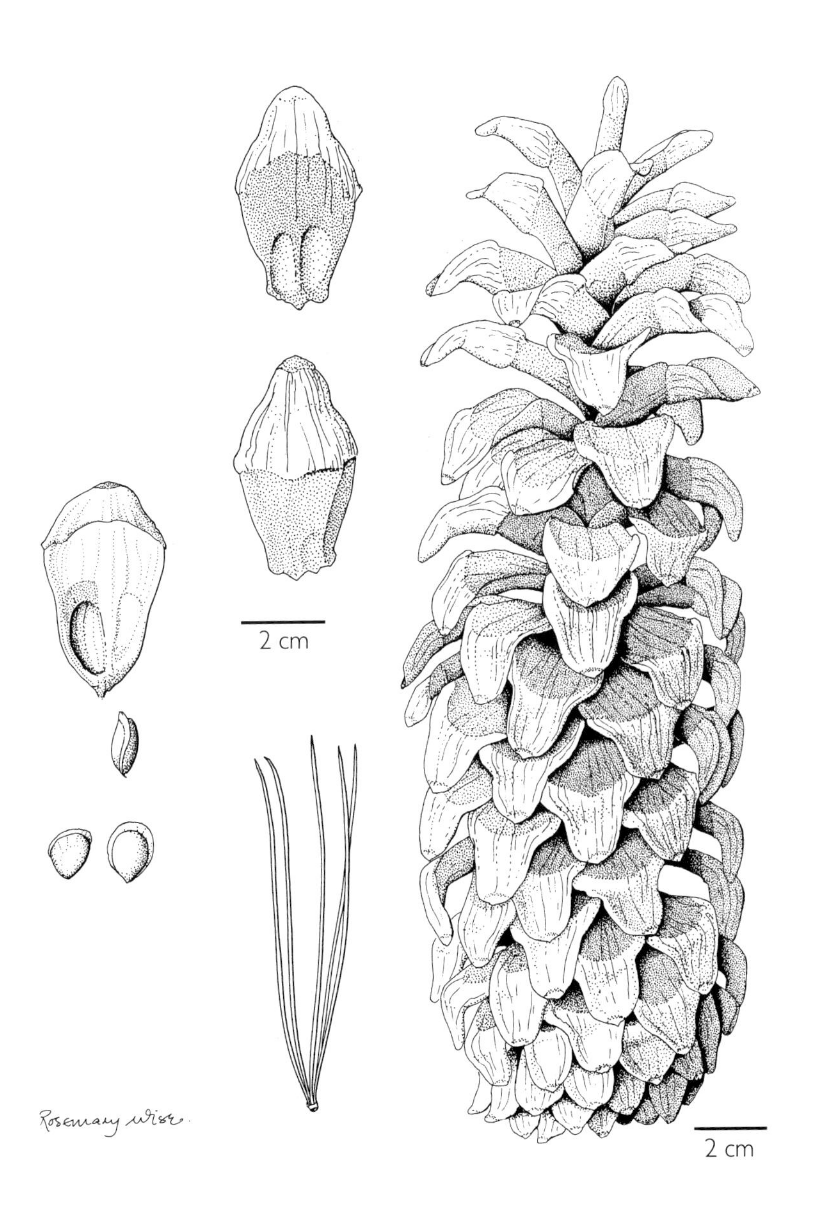
2 cm
2 cm
Rosemary Wise

38 **Pinus strobus** var. **chiapensis**

Pinus strobus Linnaeus var. *chiapensis* Martínez

Synonyms: *P. chiapensis* (Martínez) Andresen

Local names: Pinabete, Pino blanco

Habit, trunk: tree to 30–35(–40) m tall, d.b.h. to 100–150 cm, trunk straight, sometimes forked.

Bark: rough and scaly on the trunk, flaking, with shallow fissures, greyish-brown to grey; on young trees smooth, greyish-green.

Foliage twigs: slender, young shoots slightly pubescent, with small, soon eroding leaf bases (fascicle bases), grey-green; needle fascicles spreading but lax, persisting 2–3 years.

Needles: in fascicles of 5, (5–)6–12(–13) cm long, 0.6–0.8(–1) mm wide, straight or slightly twisted, lax, with (sometimes weakly) serrulate margins; stomata only on the two inner sides; scales of fascicle sheaths soon falling.

Cones: usually in whorls of 2–4 on slender peduncles, pendulous, deciduous, cylindrical to ovoid-oblong when open, (6–)8–16(–25) × 4–8 cm when open, dull brown.

Cone scales: 40–100, opening soon, thin and flexible; apophysis thin, rhombic, obtuse, straight (the basal scales not recurved), with an obtuse terminal umbo, often resinous.

Seeds: 7–9 × 4–5 mm, brown, with an adnate wing 20–30 × 6–9 mm.

Habitat: subtropical to warm temperate montane forests with high rainfall and fog (cloud forest), on well drained, loamy soil. This pine is most often associated with broadleaved trees, but also with other pines.

Distribution: MEXICO: Guerrero, E Puebla, Veracruz, Oaxaca and Chiapas; GUATEMALA: El Quiche and Huehuetenango. Mostly restricted to small remnant stands.

Altitude: (500–)800–2000(–2200) m.

Related species, subspecies or varieties: Some botanists regard *P. strobus* var. *chiapensis* as a distinct species, mainly because it occurs more than 3000 km from *P. strobus* var. *strobus* in eastern North America in a very different climate. There are no discontinuous and fixed characters that distinguish the two, so we maintain its status as a variety (Farjon & Styles, Flora Neotropica Monograph 75). It differs from *P. ayacahuite* (No. 34) in its cone scales, which have short, not recurved or reflexed apophyses; the cones are usually also smaller, but very variable and the largest cones of *P. strobus* var. *chiapensis* may be bigger than the smallest cones of *P. ayacahuite*.

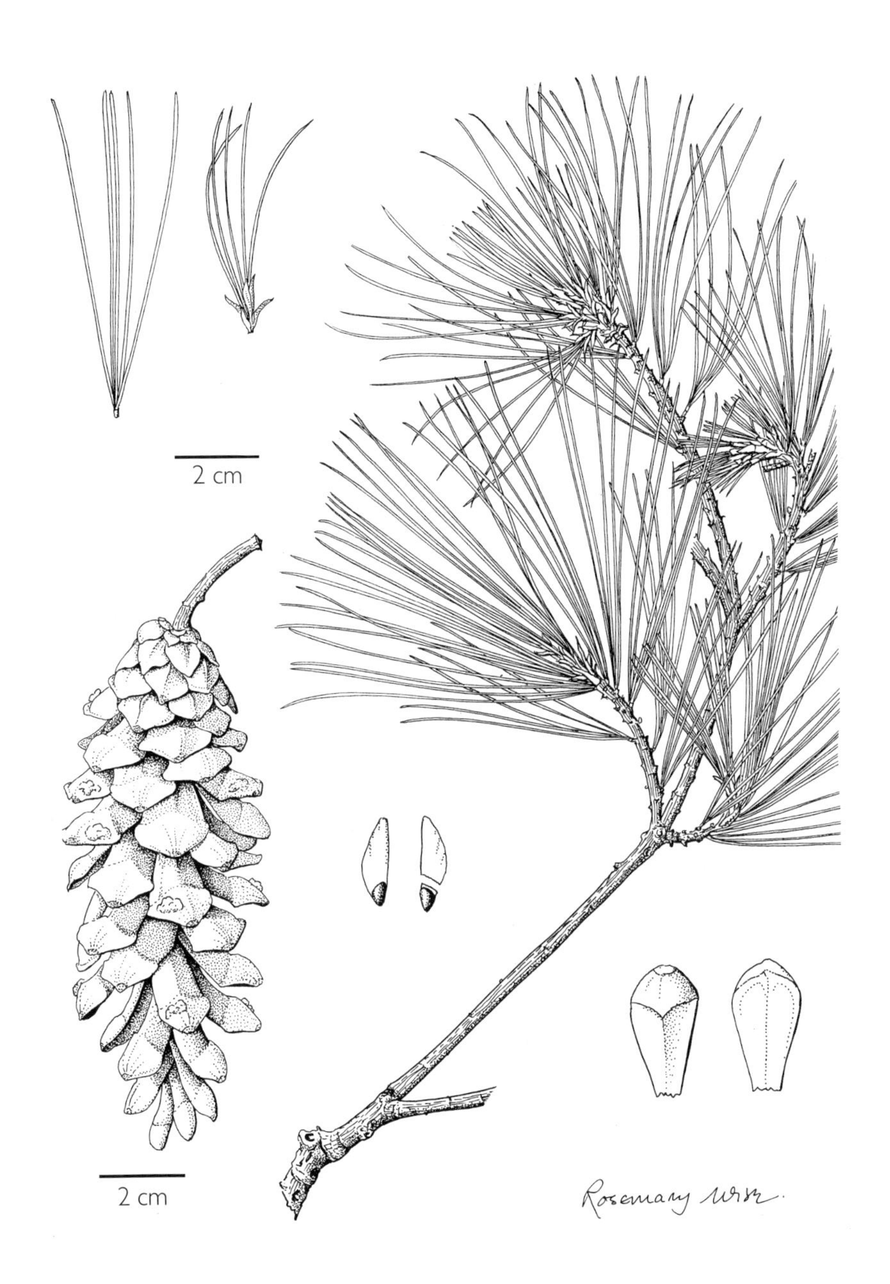
2 cm
2 cm
Rosemary Wise

39 Pinus rzedowskii

Pinus rzedowskii Madrigal Sanchez & Caballero Deloya

Local names: Ocote, Pino

Habit, trunk: tree to 15–30 m tall, d.b.h. 30–60 cm, trunk straight or curved, often branching low.

Bark: thick on the trunk, scaly, breaking into large, loose plates divided by deep, longitudinal fissures, dark brown; on young trees soon scaly and flaking, reddish-brown or grey-brown.

Foliage twigs: slender, smooth, with small leaf bases (fascicle bases), grey; needle fascicles spreading but very lax, persisting 2–3 years.

Needles: in fascicles of (3–)4–5, straight or slightly reflexed, 6–10 cm long, 0.6–0.8 mm wide, margins serrulate, stomata only on the two inner sides; scales of fascicle sheaths recoiling before they fall.

Cones: solitary or in whorls of 2–4 on slender, curved peduncles, pendulous or spreading, deciduous, ovoid-oblong, 10–15 × 6–8.5 cm when open.

Cone scales: 80–120, opening soon, thin, more or less flexible, straight; apophysis prominent, transversely keeled, with a raised umbo, light brown, often resinous.

Seeds: (6–)8(–10) × (4–)5–6 mm, dark brown, with an articulate wing 20–30(–35) × 8–13 mm.

Habitat: limited areas with rocky slopes or limestone boulders in montane pine forest, in two localities without other pine species, in the third known locality with a few other pines. This species seems restricted to places with limestone rock.

Distribution: MEXICO: Michoacán, in three separate localities in the district of Coalcomán, near Dos Aguas and 40 km W of Dos Aguas.

Altitude: 2100–2400 m.

Related or similar species: This remarkable species, discovered in 1966, has a unique combination of characters placing it "halfway" between the Piñon pines and the White pines (e.g. *P. ayacahuite* No. 34). It has also characters of the pines in subgenus *Pinus*, such as the articulate seed wings and the scaly, thick bark. It is possibly the rarest pine in Mexico.

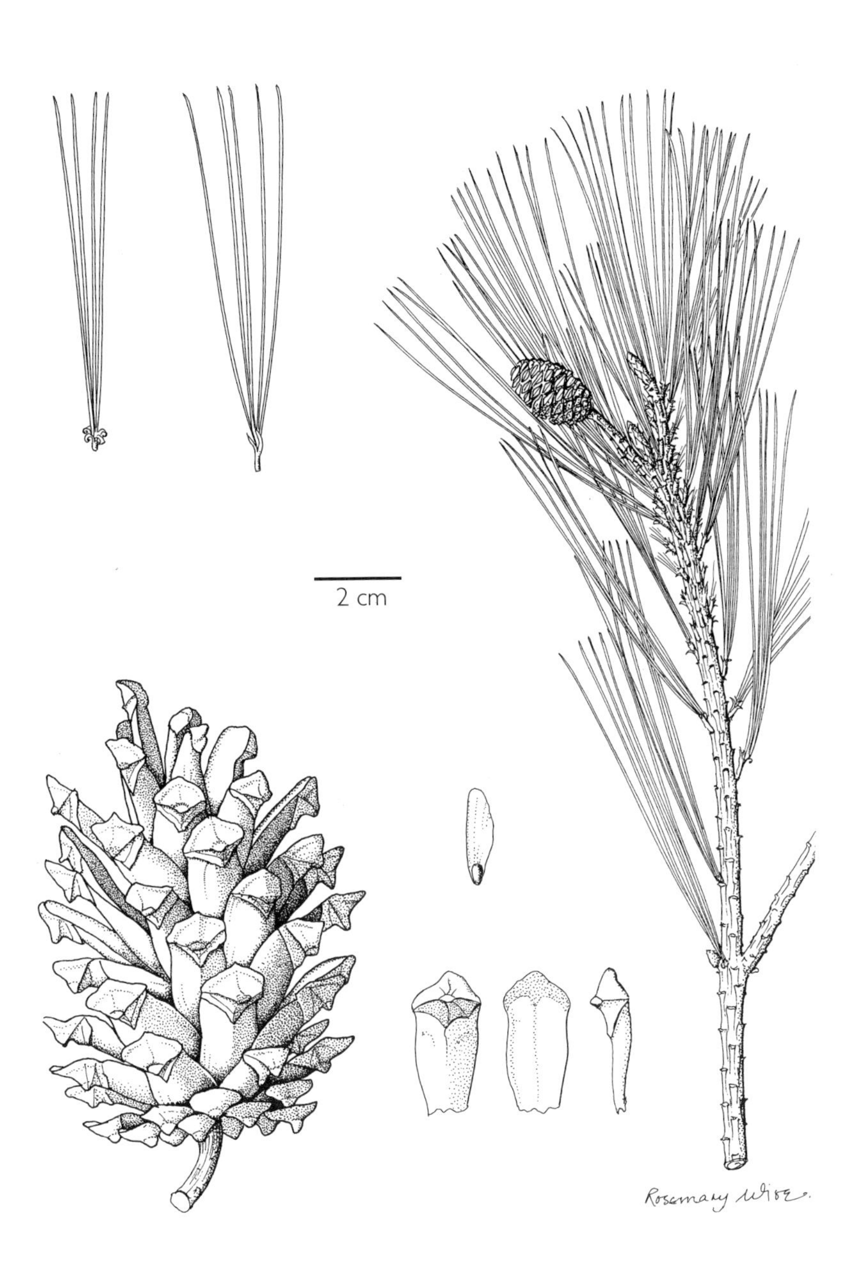
2 cm
Rosemary Wise

40 Pinus maximartinezii

Pinus maximartinezii Rzedowski

Local names: Piñón, Piñón real

Habit, trunk: tree to 5–10(–15) m tall, d.b.h. to 40–50 cm, trunk short, straight or curved, branches very long and ascending or erect.

Bark: thick on the lower part of the trunk, smooth on young trees but on larger trunks breaking into small, rough plates, inner bark orange-brown, outer bark grey.

Foliage twigs: slender, slow growing, mostly glabrous, with small leaf bases (fascicle bases), young shoots glaucous, turning orange-brown to grey; needle fascicles spreading, lax, persisting 2 years.

Needles: in fascicles of 5, rarely 3 or 4, 7–11(–13) cm long, 0.5–0.7 mm wide, with entire margins and stomata only on the two inner sides, straight, lax, usually glaucous green; scales of fascicle sheaths recoiling before they fall.

Cones: solitary, on a short peduncle, very large and heavy when mature, pendulous on slender branches, deciduous, ovoid-truncate, (15–)17–25 (–27) × 10–15 cm when open, green until the second year, then turning brown.

Cone scales: 80–110, opening slowly and only partially, thick woody, with deep seed cavities; apophysis very thick woody, raised and transversely keeled, on the lower scales conical, curved, with large, obtuse umbos.

Seeds: oblong, 20–28 × (8–)10–12 × 7–10 mm, with a hard, thick seed coat, usually remaining lodged between the seed scales; no wing attached to the seed.

Habitat: open, dry pine and pine-oak woodland, in shallow soil on calcareous sandstone and metamorphic rock, on slopes and in small canyons, with very few other pines.

Distribution: MEXICO: S Zacatecas, in the southern part of the Sierra de Morone SW of Juchipila.

Altitude: 1800–2400 m.

Notes: This remarkable pine was discovered by Dr. Jerzy Rzedowski on the market of Juchipila, where its seeds were sold as piñon nuts, because he realized that the seeds are much larger than of any species then known.

Related species: This very rare species is only distantly related to the other Piñon pines, especially *P. cembroides* (No. 43), and deserves protection from fire and livestock grazing as well as cutting. Its massive cones, but also its very slender, relatively long needles in fascicles of mainly 5 distinguish it from other Piñon pines.

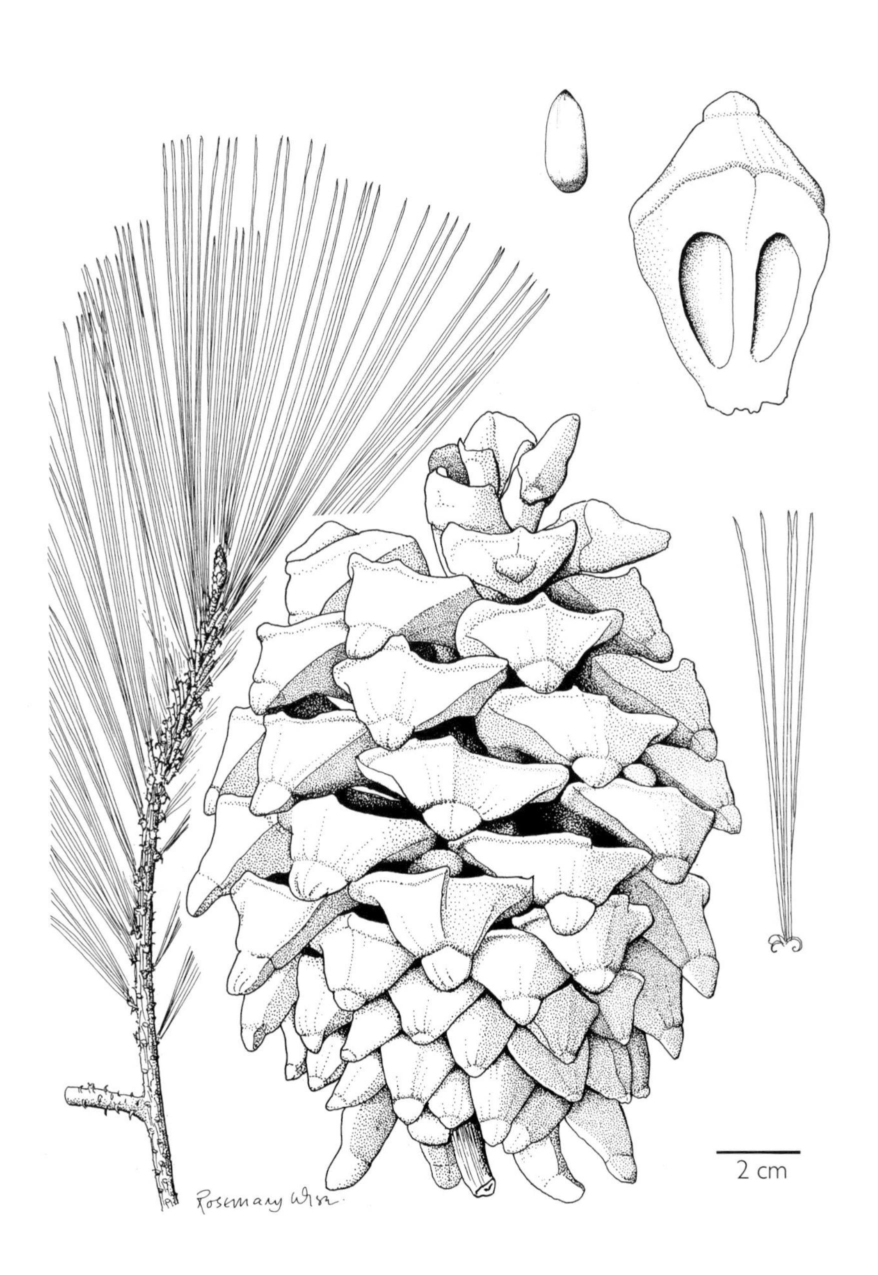
2 cm
Rosemary Wise

41 Pinus nelsonii

Pinus nelsonii Shaw

Local names: Piñón, Piñón prieto

Habit, trunk: small tree up to 5–10 m tall, d.b.h. to 15–30 cm, trunk straight, sometimes branching low, branches nearly erect or spreading.

Bark: thin, smooth, only on the lower part of largest trunks scaly with thin, small plates, ash-grey, with brown areas, on young trees almost white.

Foliage twigs: slender, stiff, erect, smooth, with small leaf bases (fascicle bases), sometimes glaucous, soon greyish-white; needle fascicles spreading or erect, persisting 2–3 years.

Needles: in fascicles of 3, rarely 4, connate, not parting until shortly before the fascicle falls (therefore seemingly single-needled), 4–8(–10) cm long, 0.7–0.8 mm wide, straight, curved or twisted, usually dark green; fascicle sheaths persistent.

Cones: rare, often on the top shoot of young trees, solitary or in pairs on very long, recurved, persistent peduncles (cones fall without them), oblong, irregular, (5–)7–12 × 4–5.5 cm when open, dark reddish-brown, more or less continuously growing.

Cone scales: 60–100, opening slightly, weakly attached to the axis, with deep seed cavities containing the seeds; apophysis thick, irregular, raised and transversely keeled, becoming wrinkled, with an obtuse, sometimes spined umbo.

Seeds: 12–15 × 8–10 mm, often only one develops fully on each scale, with a hard seed coat; no wing attached to the seed.

Habitat: semi-arid foothills and mesas of the Sierra Madre Oriental, above desert vegetation and below or just inside the "Piñon belt" with *P. cembroides*, mostly on limestone outcrops with shallow soils.

Distribution: MEXICO: S Nuevo León, W Tamaulipas, San Luis Potosí.

Altitude: 1600–2300(–2450) m.

Related species: This unique pine is only distantly related to the other "Piñon pines" and has some characters shared by none of them. Its whitish bark, connate needles with persistent fascicle sheaths and long peduncled, red-brown cones which fall without the peduncles easily distinguishes it from other Piñons.

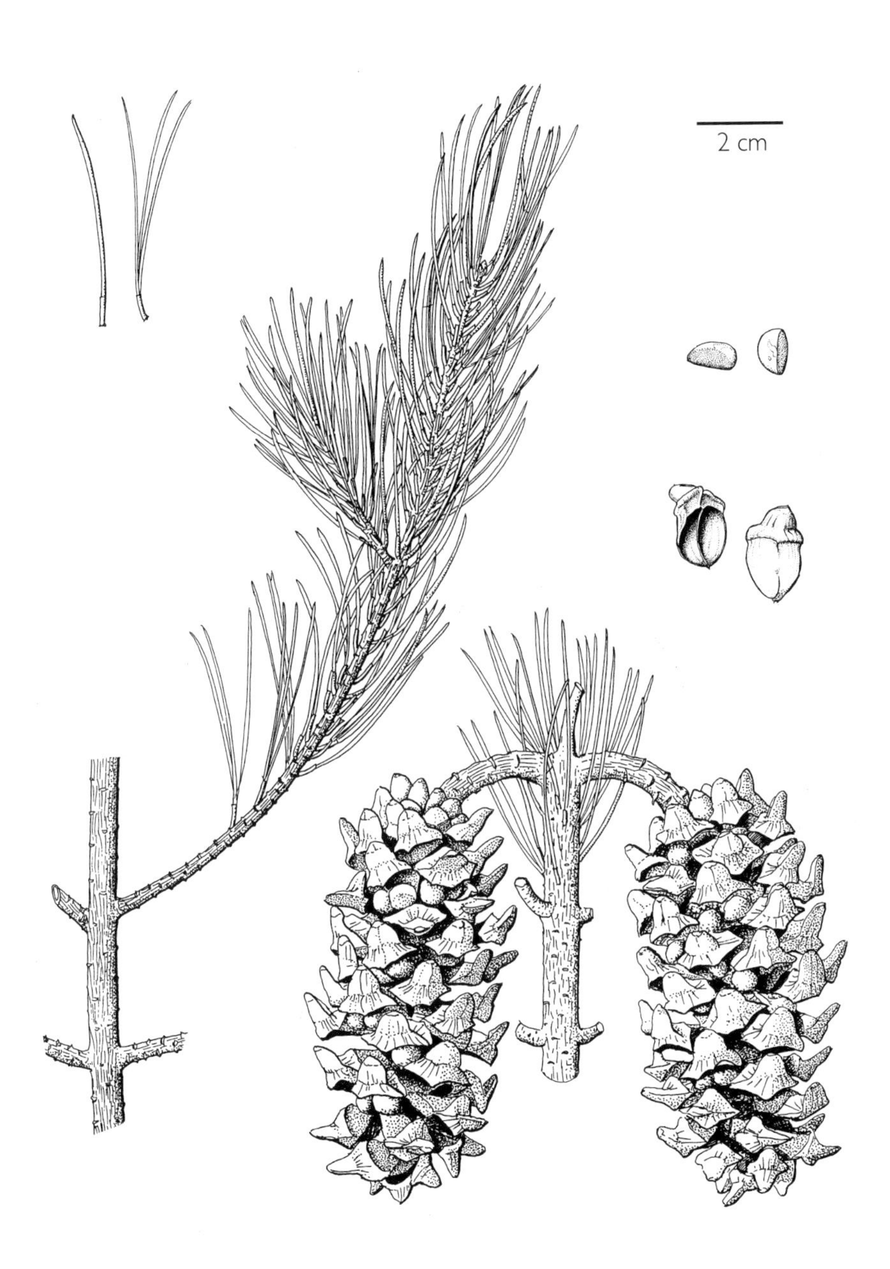
2 cm

42 Pinus pinceana

Pinus pinceana G. Gordon

Local names: Piñón, Piñón blanco

Habit, trunk: large shrub or small tree to 6–10(–12) m tall, d.b.h. to 20–30 cm, trunk short, often branching from near the ground.

Bark: thin, smooth, only on the lower part of the trunk becoming fissured and breaking into irregular plates, brownish-grey.

Foliage twigs: slender, flexible, pendulous, smooth, with small leaf bases (fascicle bases), brownish-grey to grey; needle fascicles spreading, persisting 2–3 years.

Needles: in fascicles of 3, rarely 4, 5–12(–14) cm long, 0.8–1.2 mm wide, straight, rigid, with entire margins, stomata mostly on the two inner sides, greyish-green or light green; fascicle sheaths deciduous.

Cones: solitary or rarely in pairs on slender, curved and easily breaking peduncles, pendulous, deciduous, ovoid-oblong, irregular, 5–10 × 3.5–6(–7) cm when open, lustrous red-brown.

Cone scales: 60–80, opening slightly, weakly attached to the axis and easily moveable, with deep seed cavities containing the seeds; apophysis prominently raised, transversely keeled, smooth, lustrous brown with a flattened umbo.

Seeds: 11–14 × 7–8 mm, with a hard seed coat, often only one seed fully developed on each scale; no wing attached to the seed.

Habitat: semi-arid mountains, often calcareous slopes and barrancas, above the desert vegetation and below or just in the "Piñon pine belt" with *P. cembroides*.

Distribution: MEXICO: scattered in Coahuila, N Zacatecas, San Luis Potosí, Querétaro and Hidalgo.

Altitude: 1400–2300 m.

Related or similar species: This species is somewhat similar to *P. nelsonii* (No. 41) in habit, but it has pendulous, not upright twigs, the needles are free, not connate, and the apophyses of the cone scales are smooth and lustrous, not wrinkled and dull. It differs from *P. cembroides* (No. 43) in its pendulous habit, its longer needles and its larger, longer and pendulous cones, of which the scales only partly open.

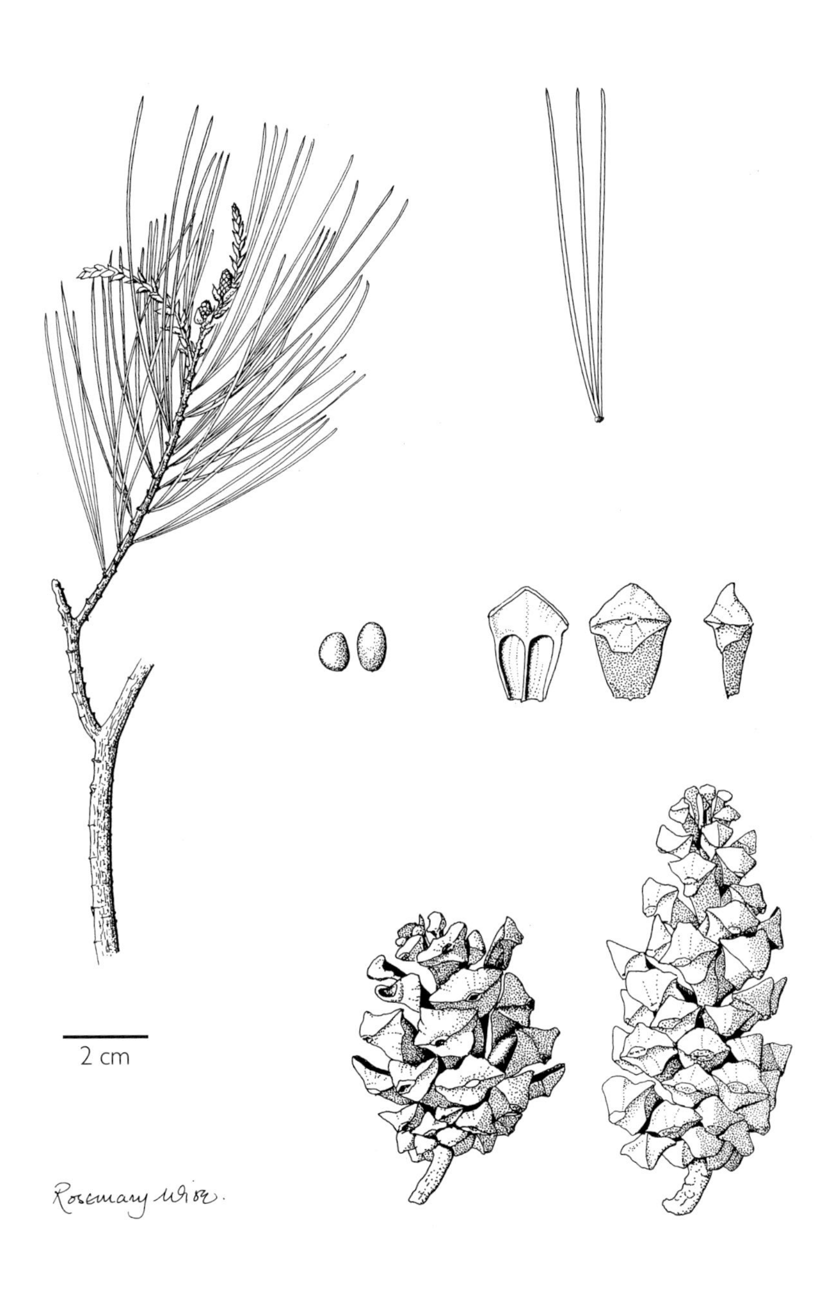
2 cm
Rosemary Wise.

43 Pinus cembroides

Pinus cembroides Zuccarini subsp. *cembroides* var. *cembroides*

Local names: Piñón, Piñón prieto, Piñonero

Habit, trunk: small tree to 10–15 m, d.b.h. to 60–80 cm, trunk short, straight or contorted, branching low.

Bark: thick on the trunk, scaly, breaking into small, irregular plates and longitudinal fissures, inner bark yellowish-orange, outer bark dark grey.

Foliage twigs: slender, with small leaf bases (fascicle bases), smooth, orange-brown or glaucous, turning grey; needle fascicles spreading or erect, persisting 3–4 years.

Needles: in fascicles of 2–3, curved or nearly straight, (2–)3–5(–6.5) cm long, (0.6–)0.7–1 mm wide, with entire margins and stomata on all sides, green on the abaxial side and greyish on the adaxial faces; scales of fascicle sheaths recoiling before they fall.

Cones: solitairy or in whorls of 2–3 on short peduncles, deciduous, usually wider than long when open, irregular, 2–5 × 3–6 cm when open.

Cone scales: 25–40, only 10–15 fertile, thin, flexible, with deep seed cavities containing the seeds, spreading very wide; apophysis raised, transversely keeled, with a flat or raised umbo, light brown or reddish-brown.

Seeds: 10–13 × 6–10 mm, with a hard seed coat, usually only one seed develops on a scale; no wing attached to the seed.

Habitat: semi-arid foothills and plateaus in a zone between semi-desert vegetation below and mesic pine forest above, usually on very thin, rocky soils. This species is commonly associated with *Juniperus*.

Distribution: MEXICO: NE Sonora, Chihuahua, Coahuila, Durango, Zacatecas, Nuevo León, W Tamaulipas, San Luis Potosí, Aguascalientes, NE Jalisco, N Guanajuato, Querétaro, Hidalgo, México, D.F., Tlaxcala, Veracruz and Puebla; also in SW U.S.A.

Altitude: (800–)1500–2600(–2800) m.

Related species, subspecies and varieties: This variable species has a number of subspecies and varieties, which have also been recognized as distinct species. *Pinus cembroides* subsp. *cembroides* var. *bicolor* Little (syn.: *P. discolor* D. K. Bailey & F. G. Hawksworth; *P. johannis* Robert-Passini) is a shrub or a small tree with needles in fascicles of 3, rarely 2, 4 or 5, with stomata only on the two whitish inner sides. It is found scattered in NE Mexico and SW U.S.A. *Pinus cembroides* subsp. *lagunae* (Robert-Passini) D. K. Bailey (syn.: *P. lagunae* (Robert-Passini) Passini) has longer (but variable) needles than the other varieties (up to 8 cm); it occurs only in the Sierra de la Laguna of Baja California Sur. *Pinus cembroides* subsp. *orizabensis* D. K. Bailey (syn.: *P. orizabensis* (D. K. Bailey) D. K. Bailey) has needles in fascicles of 3–4, rarely 5 and larger (but variable) cones. It is found in Puebla and adjacent parts of Tlaxcala and Veracruz. All these variants have thick, hard seed coats and pinkish macrogametophytes ("endosperm"), which distinguish the species as a whole from closely related species of Piñón, except *P. culminicola* (No. 44).

2 cm
Rosemary Wise.

44 Pinus culminicola

Pinus culminicola Andresen & Beaman

Local names: Piñón

Habit, trunk: shrub, usually decumbent, 1–5 m tall, multistemmed, diameter of stems to 15–25 cm, often forming an extensive, dense vegetation ("matorral").

Bark: thin, scaly, with small, irregular plates, grey-brown.

Foliage twigs: short, thick, with short decurrent, persistent leaf bases (fascicle bases), flexible, erect; needle fascicles spreading to erect, persisting 2–3 years.

Needles: in fascicles of 5 (very rarely 4–6), 3–5 cm long, 0.9–1.3 mm wide, curved, rigid, margins nearly entire; stomata only on the two inner sides, in whitish bands; scales of fascicle sheaths recoiling before they fall.

Cones: solitary or in pairs on short peduncles, deciduous, usually wider than long when open, then 3–4.5 × 3–5 cm.

Cone scales: 45–60, only 10–20 fertile, irregular, thin, flexible, with deep seed cavities containing the seeds, spreading very wide; apophysis slightly raised, transversely keeled, yellowish-brown, with a darker umbo, often resinous.

Seeds: 5–7 × 4–5 mm, with a thick, hard seed coat (often only 1 seed develops on a scale); no wing attached to the seed.

Habitat: mountain summit areas and ridges, on wind-exposed sites with shallow, rocky and calcareous soils. Often, but not everywhere, this species grows in extensive, dense stands forming "matorral" with no other pines present; in some places it occurs in open forest.

Distribution: MEXICO: SE Coahuila, S-Central Nuevo León, on mountain summits; the largest population is on Cerro Potosí.

Altitude: 3000–3700 m.

Related species: *Pinus culminicola* is related to *P. cembroides* (No. 43) and its subspecies and varieties. It differs obviously in its low, shrubby habit with many short, curved stems. It has also a higher number of needles per fascicle than the other taxa. Its cones and seeds are very similar.

2 cm

45 Pinus remota

Pinus remota (Little) D. K. Bailey & F. G. Hawksworth

Synonyms: *P. cembroides* var. *remota* Little; *P. catarinae* Robert-Passini

Local names: Piñón, Piñonero

Habit, trunk: shrub to small tree 3–9 m tall, d.b.h. 15–40 cm, trunk short, contorted, branching low.

Bark: thick on the trunk, scaly, on the lower part of the trunk longitudinally furrowed, grey to blackish-grey.

Foliage twigs: slender but rigid, with prominent but small leaf bases (fascicle bases) leaving circular scars; needle fascicles spreading, persisting 4–5 years.

Needles: in fascicles of 2(–3), curved, (2–)3–4.5(–5.5) cm long, 0.8–1.1 mm wide, margins entire, stomata on all sides; fascicle sheaths soon deciduous, the scales not recoiling to a rosette.

Cones: solitary or in pairs on short peduncles, deciduous, opening wide, with a flat base, wider than long (2–)2.5–4 × 3–6 cm when open.

Cone scales: 25–35, only ca. 10 are fertile, with deep seed cavities containing the seeds, irregular, spreading wide; apophysis raised, (transversely) keeled, light brown or red-brown, with a flat, darker umbo.

Seeds: 12–16 × 8–10 mm, with a very thin, weak seed coat (often only one seed develops per scale); no seed wing attached to the seed.

Habitat: canyons and mountain slopes, usually of limestone, with very shallow and rocky soil, where Pinyon-Juniper woodland is poorly developed; often in vegetation with *Opuntia* and *Agave*.

Distribution: MEXICO: NE and SE Chihuahua, Coahuila and W Nuevo León, in widely separated stands; also in W Texas (U.S.A.).

Altitude: 1200–1600(–1850) m.

Related species, subspecies or varieties: In the U.S.A., this species has been described as a variety of *P. cembroides* (No. 43) or even as a synonym of it. However, it is distinct from that species especially in the seeds, which have a very thin seed coat (easily crushed) and a white macrogametophyte ("endosperm"). It resembles *P. edulis*, a species which occurs only in the U.S.A., with predominantly 2 needles per fascicle and similar seeds. *Pinus edulis* has thicker needles (1.0–1.4 mm) with only two resin ducts (a microscopic character not visible in the field) instead of 2–5; its fascicle sheaths recoil to a rosette before falling, and the apophysis on the cone scales is more prominently raised.

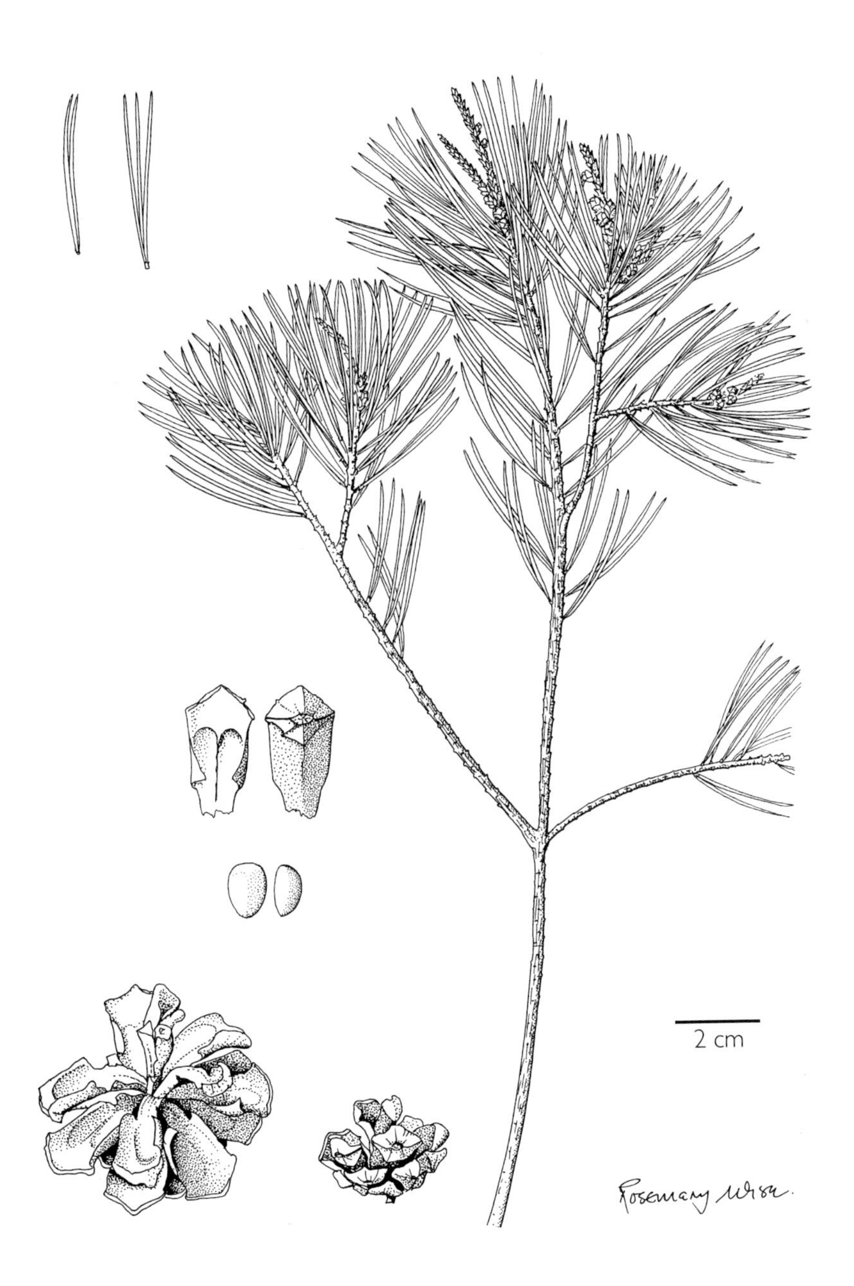
2 cm
Rosemary Wise.

46 Pinus monophylla

Pinus monophylla J. Torrey & Frémont

Synonyms: *P. californiarum* D. K. Bailey; *P. edulis* Engelmann var. *fallax* Little

Local names: Piñón

Habit, trunk: large shrub or tree to 15–20 m tall, d.b.h. to 40–50 cm, trunk short, straight or contorted, branching low.

Bark: thick on the trunk, scaly, with small plates and shallow, longitudinal fissures, inner bark orange, outer bark reddish-brown to grey.

Foliage twigs: short, stout, rigid, smooth, with small leaf bases (fascicle bases), orange-yellow, turning grey; needle fascicles spreading or erect, persisting 4–8 years.

Needles: in fascicles of 1, rarely 2, curved, (2–)2.5–6 cm long, 1.2–2.2(–2.5) mm wide, round, acute, often very glaucous; stomata all around in distinct lines; fascicle sheaths deciduous, the scales not recoiling to a rosette.

Cones: solitary or in whorls of 2–4 on short peduncles, deciduous, opening wide, more or less rounded at base and 4–6 × 4.5–7 cm when open.

Cone scales: 30–50, only the central 6–12 fertile, with deep seed cavities containing the seeds; apophysis thick woody, prominently raised and often conical, yellowish-brown with an obtuse or flat umbo, often very resinous.

Seeds: 13–18 × 8–12 mm, with a thin, fragile seed coat (often only one seed develops per scale); no wing attached to the seed.

Habitat: in or just above the chaparral zone, often with *P. quadrifolia* and/or *Juniperus californica*, but not forming extensive Pinyon-Juniper woodland, on rocky, shallow soils.

Distribution: MEXICO: Baja California Norte, in the Sierra Juarez, Sierra San Pedro Martír and Sierra La Asamblea; also widespread in the SW U.S.A.

Altitude: (950–)1200–1700(–2000) m.

Notes: This species is usually quickly recognized by its single needles, but ocasionally two needles are found in fascicles on the same tree.

Related species, subspecies or varieties: In Baja California, *P. monophylla* often occurs in the same area as *P. quadrifolia* (No. 47). The two species may occasionally hybridize and trees with predominantly 2–3 needles per fascicle may be of such hybrid origin. *Pinus quadrifolia* has normally (3–)4(–5) needles per fascicle. The cones of the two species are very similar.

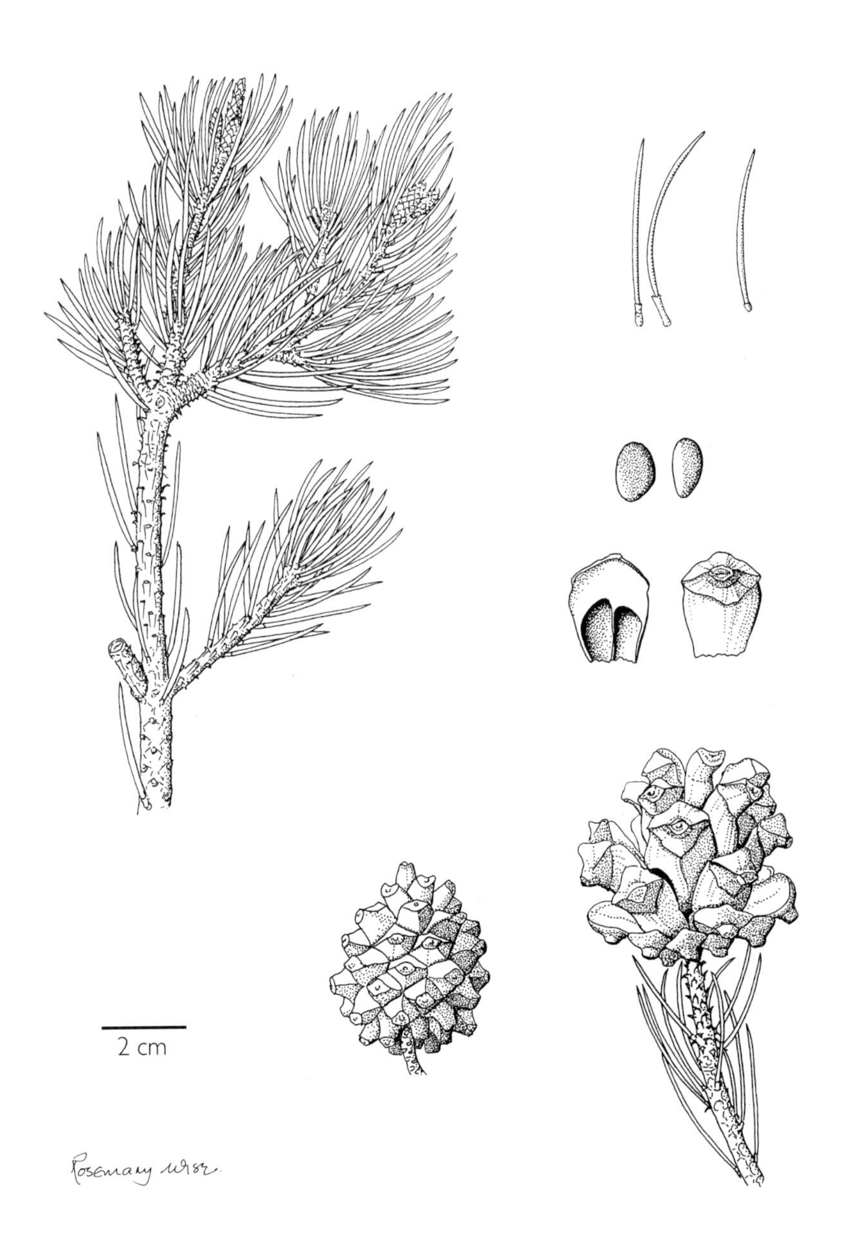
2 cm

47 Pinus quadrifolia

Pinus quadrifolia Sudworth

Synonyms: *P. cembroides* var. *parryana* (Engelmann) A. Voss; *P. juarezensis* Lanner

Local names: Piñón

Habit, trunk: large shrub or tree to 10–15 m tall, d.b.h. to 30–50 cm, trunk short, usually low branched.

Bark: thick on the trunk, scaly, with deep fissures, inner bark yellowish-orange, outer bark brown, becoming grey.

Foliage twigs: stout and rigid, with small leaf bases (fascicle bases), soon grey; needle fascicles spreading or erect, some distance apart, persisting (3–)4–7 years.

Needles: in fascicles of (3–)4(–5), rarely a few fascicles of 2 or 6, usually curved, (1.5–)2–4(–5) cm long, (0.8–)1–1.5(–1.7) mm wide, margins entire, stomata only on the two inner sides; fascicles sheaths deciduous, the scales recoiling only slightly.

Cones: solitary or in whorls of 2–4 on short peduncles, deciduous, opening wide, when open somewhat rounded at the base and 4–6 × 4.5–7 cm (often wider than long).

Cone scales: 30–50, only the central 6–12 fertile, irregular, with deep seed cavities; apophysis thick woody, pyramidal or conical, light brown or reddish-brown, with an obtuse umbo.

Seeds: 12–18 × 8–12 mm, with a thin, fragile seed coat (often only one seed per scale develops fully); no wing attached to the seed.

Habitat: mainly a zone between chaparral or semi-desert vegetation and mixed coniferous forest, on granitic or volcanic, rocky soils, in some areas forming Pinyon-Juniper woodland, with *Juniperus californica* and often *Quercus turbinella*.

Distribution: MEXICO: Baja California Norte, to ca. 30° 30' N in the foothills of the Sierra San Pedro Martír; also in Alta California in San Diego and Riverside Counties (U.S.A.).

Altitude: 900–2400(–2700) m

Notes: As in all Pinyon pines, the needles tend to fall individually as well as in entire fascicles. Needle numbers per fascicle should therefore always be counted on later fascicles, not on the oldest ones lower down the foliage twigs.

Related species, subspecies or varieties: This species is closely related to *P. edulis* (not in Mexico) and *P. monophylla* (No. 46).

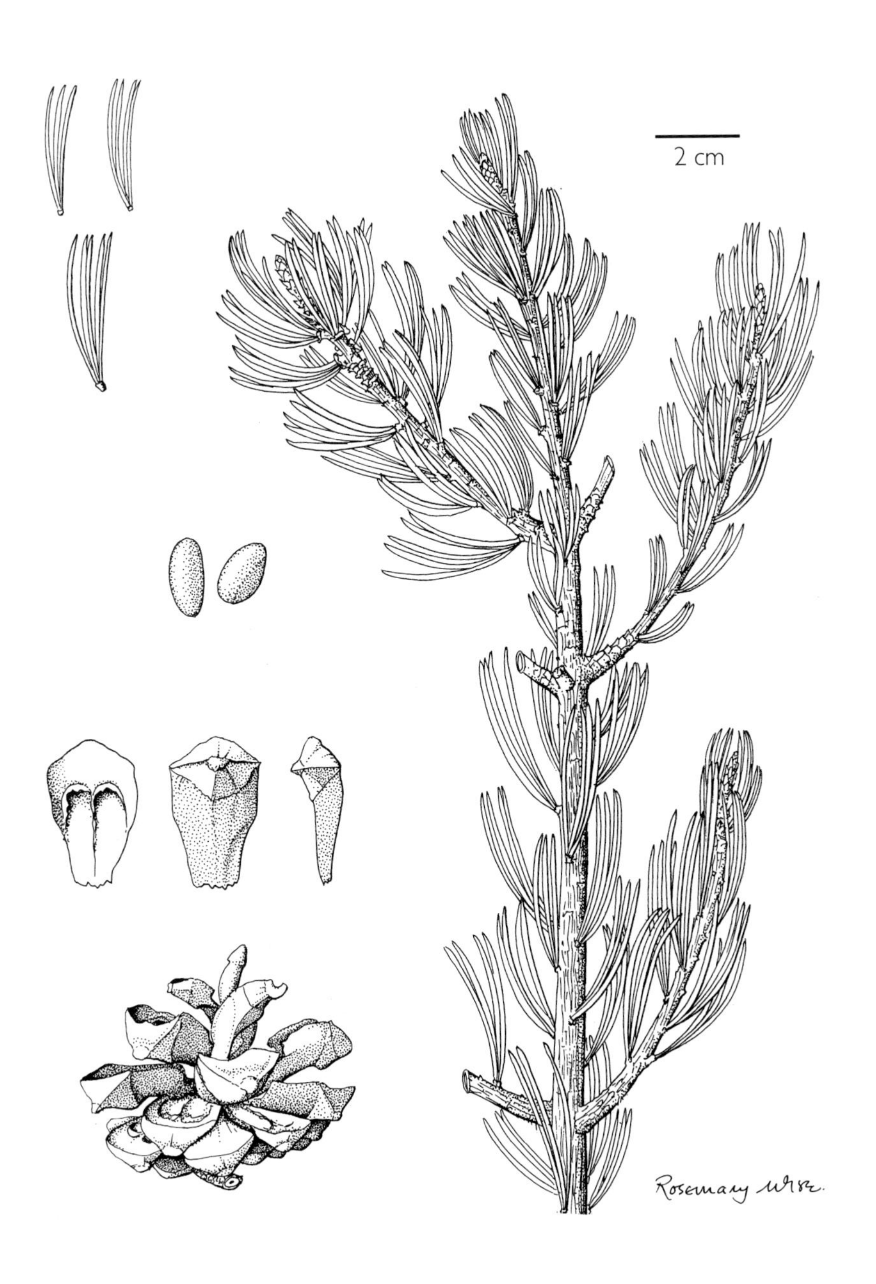
2 cm
Rosemary Wise

Literature

The references below are limited to a few recent or well known floristic works or papers and do not in any way represent an overview of the literature published on the pines of Mexico and Central America. They have not necessarily been cited in this Field Guide. Since the taxonomy and descriptions of the guide are based on the revision published nearly simultaneously in Flora Neotropica, that is obviously the most important reference in this context. However, it may be useful also to compare the descriptions given here with previous work, while the student who would be interested in further taxonomic study of the pines will find in this list of references ample information for a good introduction to the subject.

Carvajal, S. & R. McVaugh. 1992. *Pinus*. In: R. McVaugh. Flora Novo-Galiciana 17: 32–100. The University of Michigan Herbarium, Ann Arbor.

Critchfield, W. B. & E. L. Little. 1966. Geographical distribution of the pines of the world. U.S. Forest Service Miscellaneous Publication 991. Washington, D.C.

Farjon, A. 1995. Typification of *Pinus apulcensis* Lindley (Pinaceae), a misinterpreted name for a Latin American pine. Novon 5: 252–256.

Farjon, A., C. N. Page & N. Schellevis. 1993. A preliminary world list of threatened conifer taxa. Biodiversity and Conservation 2: 304–326.

Farjon, A. & B. T. Styles. 1997. *Pinus*. Flora Neotropica Monograph 75. The New York Botanical Garden, New York.

Kral, R. 1993. 6. *Pinus*. In: Flora of North America Editorial Committee (ed.). Flora of North America, North of Mexico 2: 373–398. Oxford University Press, New York.

Malusa, J. 1992. Phylogeny and biogeography of the Pinyon pines (*Pinus* subsect. *Cembroides*). Systematic Botany 17: 42–66.

Martínez, M. 1948. Los Pinos mexicanos. Ed. 2, Universidad Autónoma de México, México.

Mirov, N. T. 1967. The genus *Pinus*. Ronald Press, New York.

Perry, J. P. 1991. The Pines of Mexico and Central America. Timber Press, Portland, Oregon.

Index to species

accepted names are in **bold**, synonyms are in *italics*

Pinus arizonica var. **arizonica**, 46
Pinus arizonica var. **cooperi**, 46
Pinus arizonica var. **stormiae**, 46
Pinus attenuata, 88
Pinus ayacahuite var. **ayacahuite**, 94
Pinus ayacahuite var. *brachyptera*, 100
Pinus ayacahuite var. *novogaliciana*, 100
Pinus ayacahuite var. **veitchii**, 94
Pinus californiarum, 118
Pinus caribaea var. **hondurensis**, 42
Pinus catarinae, 116
Pinus cembroides subsp. **cembroides** var. **bicolor**, 112
Pinus cembroides subsp. **cembroides** var. **cembroides**, 112
Pinus cembroides subsp. **lagunae**, 112
Pinus cembroides subsp. **orizabensis**, 112
Pinus cembroides var. *parryana*, 120
Pinus cembroides var. *remota*, 116
Pinus chiapensis, 102
Pinus contorta var. **murrayana**, 38
Pinus coulteri, 92
Pinus culminicola, 114
Pinus devoniana, 58
Pinus discolor, 112
Pinus donnell-smithii, 52
Pinus douglasiana, 60
Pinus durangensis, 76
Pinus edulis var. *fallax*, 118
Pinus engelmannii, 50
Pinus flexilis var. **reflexa**, 98
Pinus greggii, 90
Pinus hartwegii, 52
Pinus herrerae, 40
Pinus hondurensis, 42
Pinus jaliscana, 72
Pinus jeffreyi, 48
Pinus johannis, 112
Pinus juarezensis, 120
Pinus lambertiana, 96
Pinus lagunae, 112
Pinus lawsonii, 78
Pinus leiophylla var. **chihuahuana**, 36
Pinus leiophylla var. **leiophylla**, 36
Pinus lumholtzii, 64
Pinus macvaughii, 72
Pinus martinezii, 76
Pinus maximartinezii, 106
Pinus maximinoi, 62
Pinus michoacana, 58
Pinus michoacana var. *cornuta*, 58

Pinus michoacana var. *quevedoi*, 58
Pinus montezumae var. **gordoniana**, 56
Pinus montezumae var. *lindleyi*, 56
Pinus montezumae var. *mezambrana*, 56
Pinus montezumae var. **montezumae**, 56
Pinus monophylla, 118
Pinus muricata var. *cedrosensis*, 86
Pinus muricata var. **muricata**, 84
Pinus nelsonii, 108
Pinus nubicola, 54
Pinus oocarpa var. *manzanoi*, 66
Pinus oocarpa var. *microphylla*, 68
Pinus oocarpa var. *ochoterenae*, 74
Pinus oocarpa var. **oocarpa**, 66
Pinus oocarpa var. **trifoliata**, 66
Pinus orizabensis, 112
Pinus patula var. **patula**, 70
Pinus patula var. **longipedunculata**, 70
Pinus patula subsp. *tecunumanii*, 74
Pinus pinceana, 110
Pinus ponderosa var. *arizonica*, 46
Pinus ponderosa var. **scopulorum**, 44
Pinus praetermissa, 68
Pinus pringlei, 80
Pinus pseudostrobus var. **apulcensis**, 54
Pinus pseudostrobus var. *coatepecensis*, 54
Pinus pseudostrobus var. *estevezii*, 54
Pinus pseudostrobus var. *oaxacana*, 54
Pinus pseudostrobus var. **pseudostrobus**, 54
Pinus quadrifolia, 120
Pinus radiata var. **binata**, 86
Pinus radiata var. *cedrosensis*, 86
Pinus radiata forma *guadalupensis*, 86
Pinus reflexa, 98
Pinus remorata, 84
Pinus remota, 116
Pinus rudis, 52
Pinus rzedowskii, 104
Pinus strobiformis, 100
Pinus strobus var. **chiapensis**, 102
Pinus tecunumanii, 74
Pinus tenuifolia, 62
Pinus teocote, 82
Pinus teocote var. *macrocarpa*, 82
Pinus yecorensis, 54

1. **Pinus leiophylla**

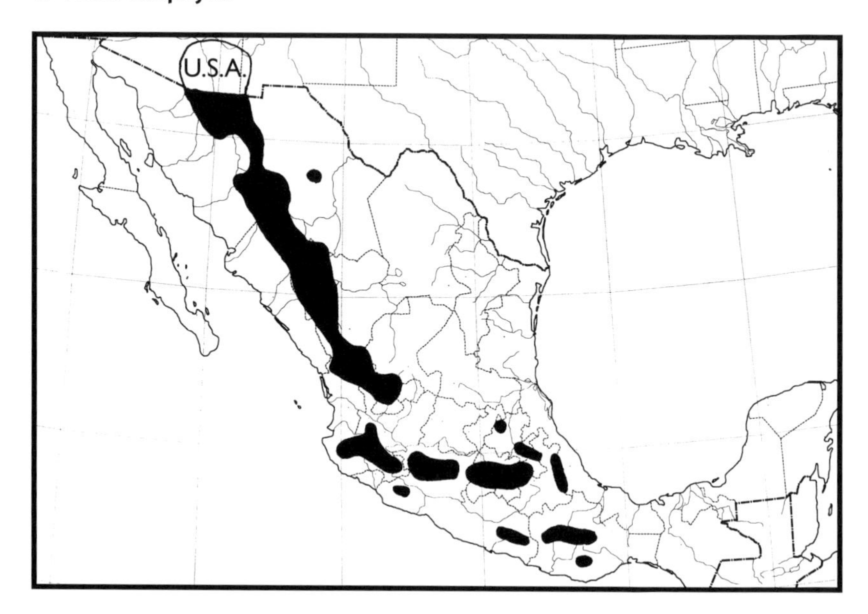

3. **Pinus contorta** var. **murrayana**

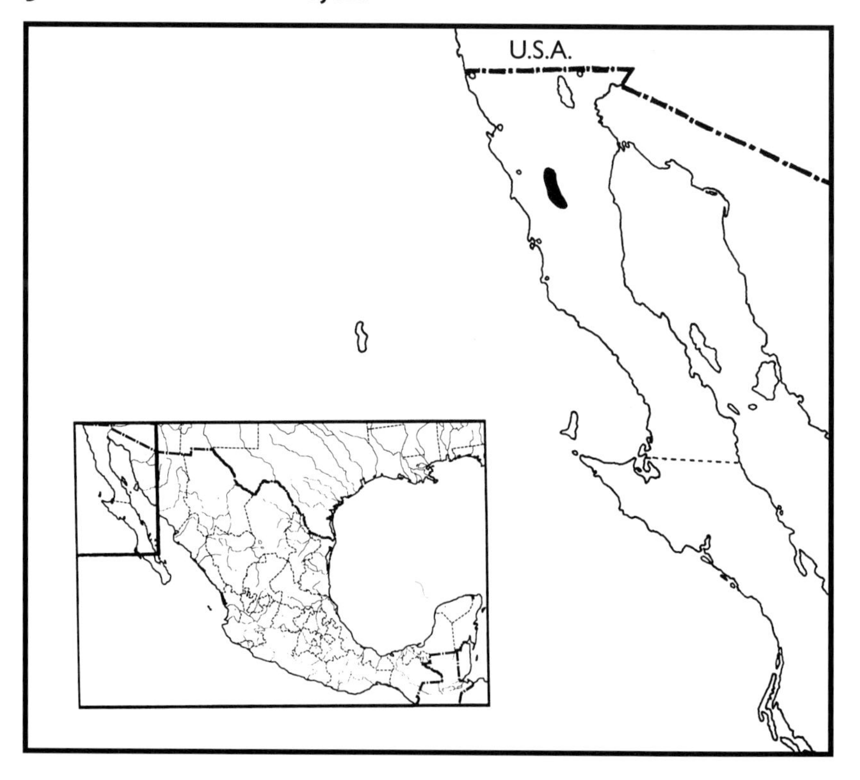

4. **Pinus herrerae**

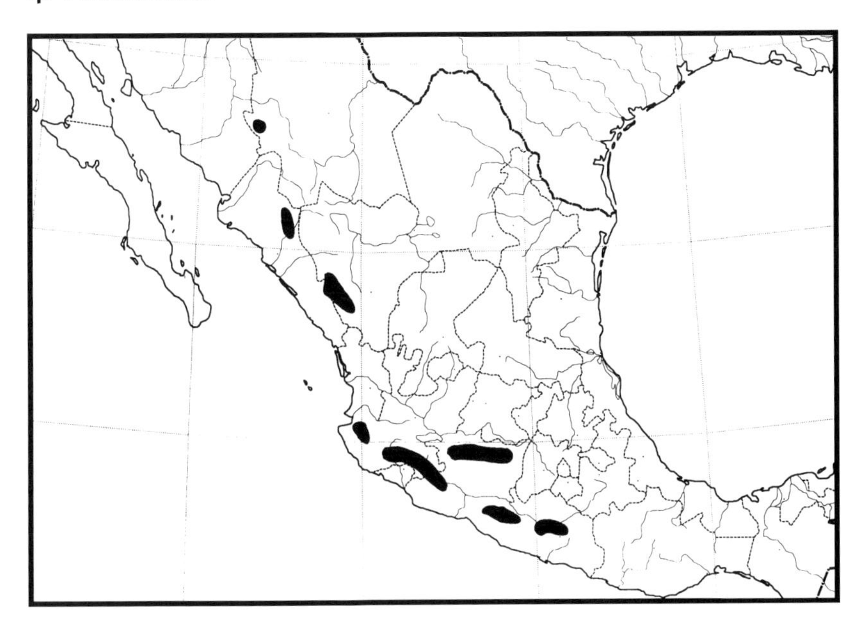

5. **Pinus caribaea** var. **hondurensis**

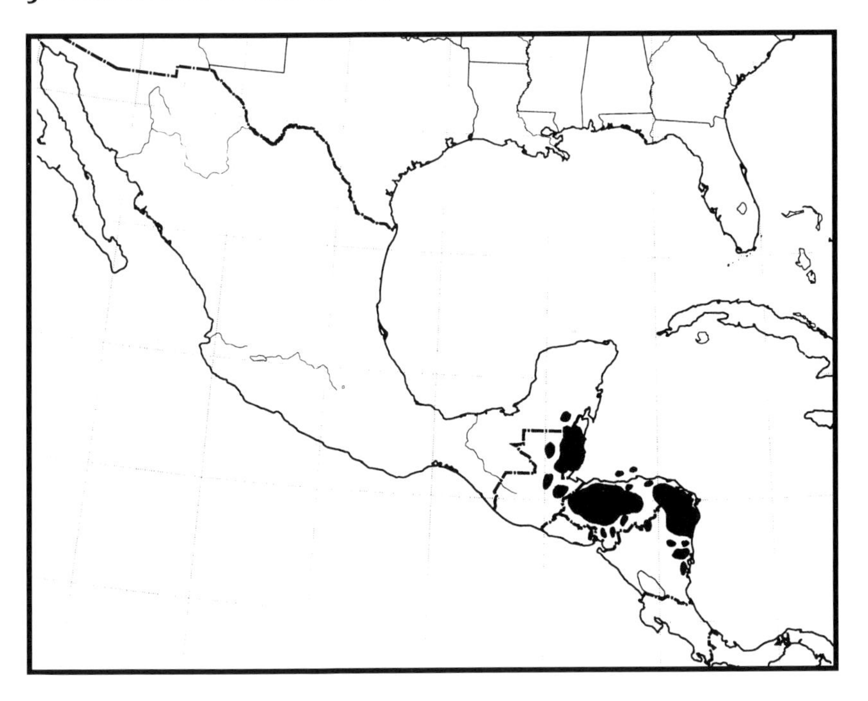

9. Pinus ponderosa var. **scopulorum**

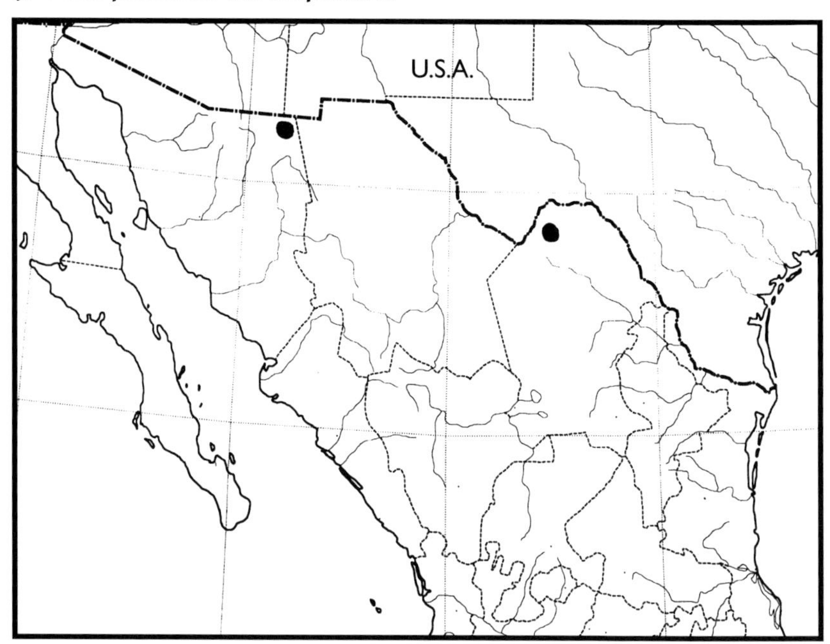

10. Pinus arizonica

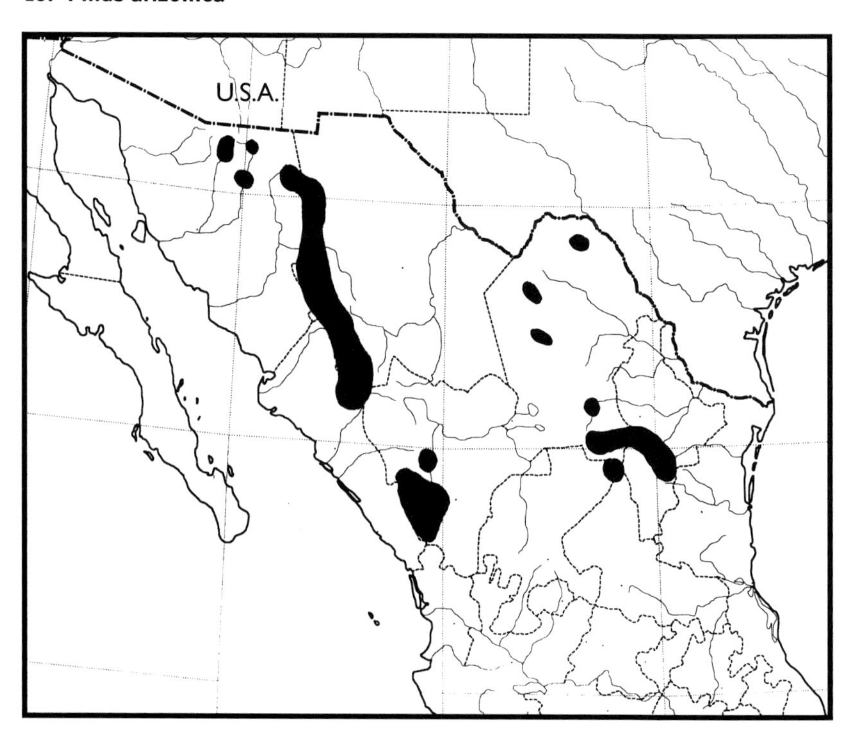

11. **Pinus jeffreyi**

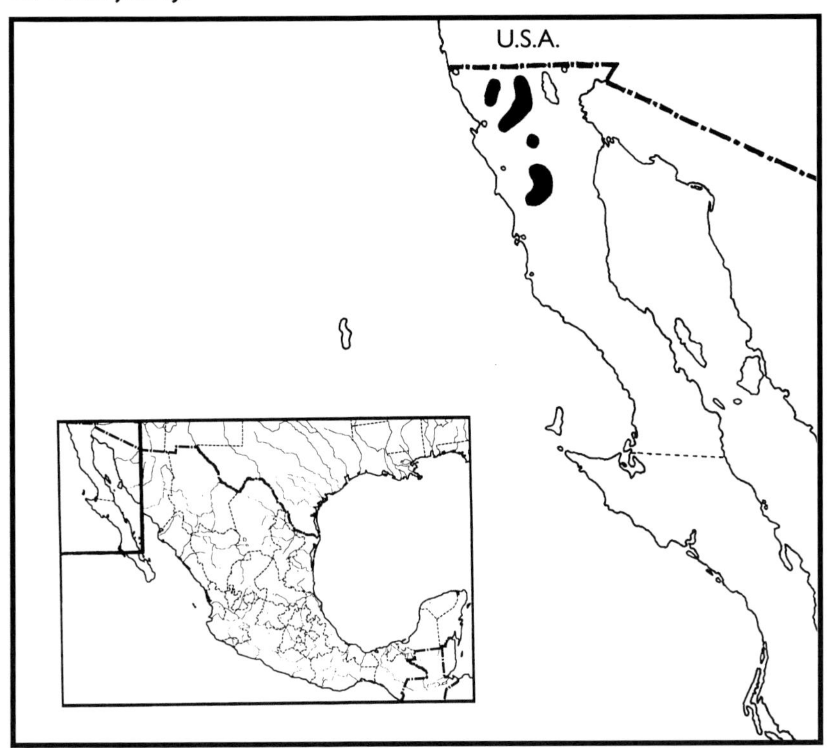

12. **Pinus engelmannii**

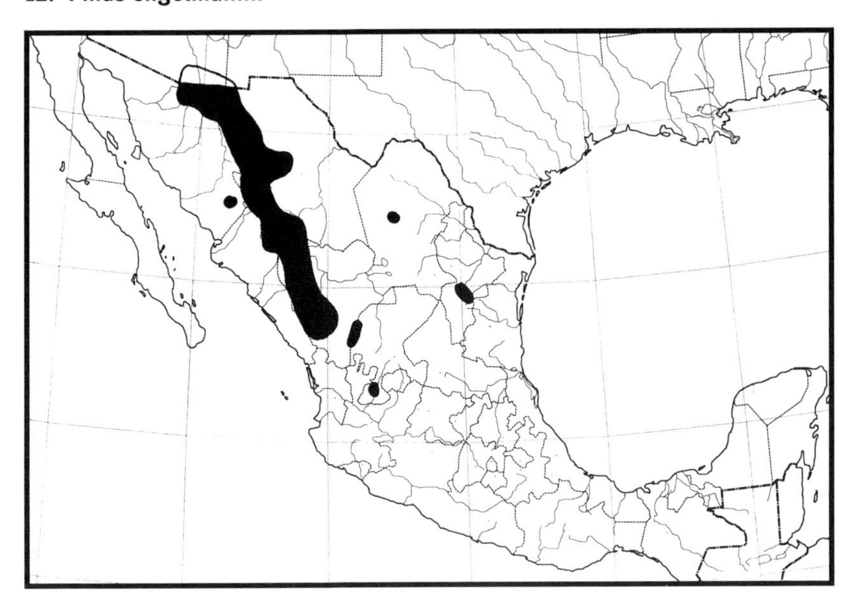

13. Pinus hartwegii

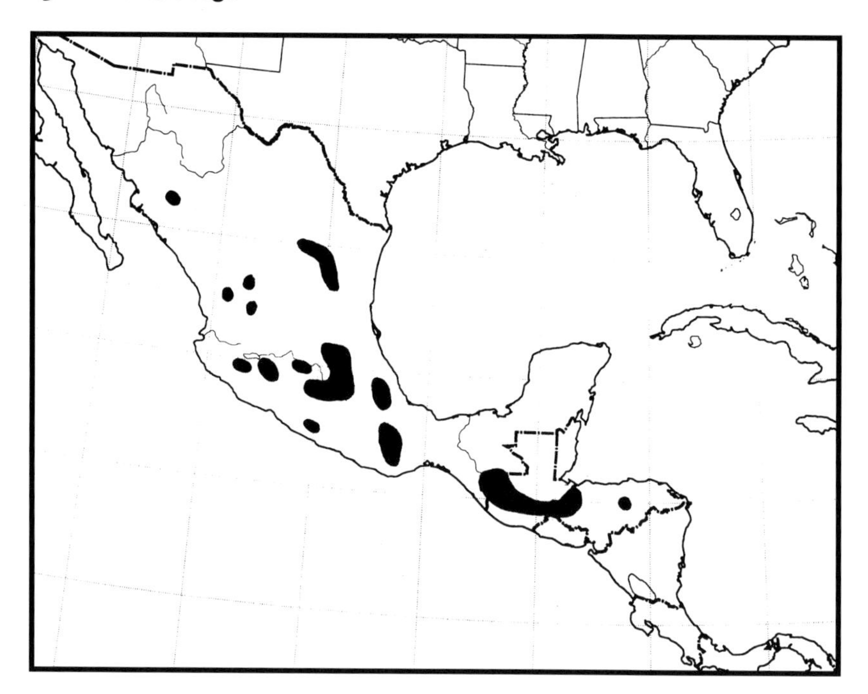

14. Pinus pseudostrobus

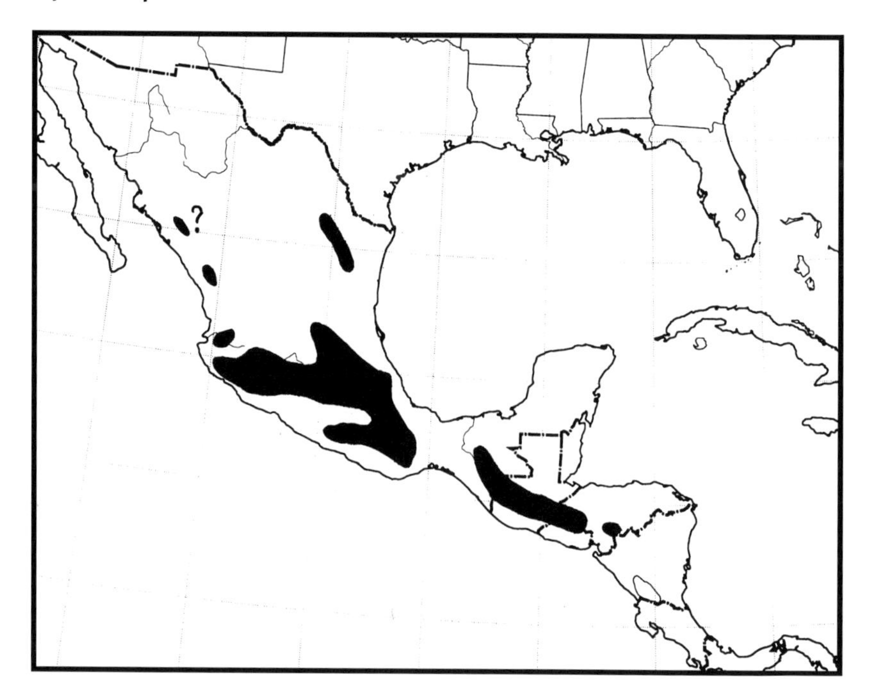

15. Pinus montezumae

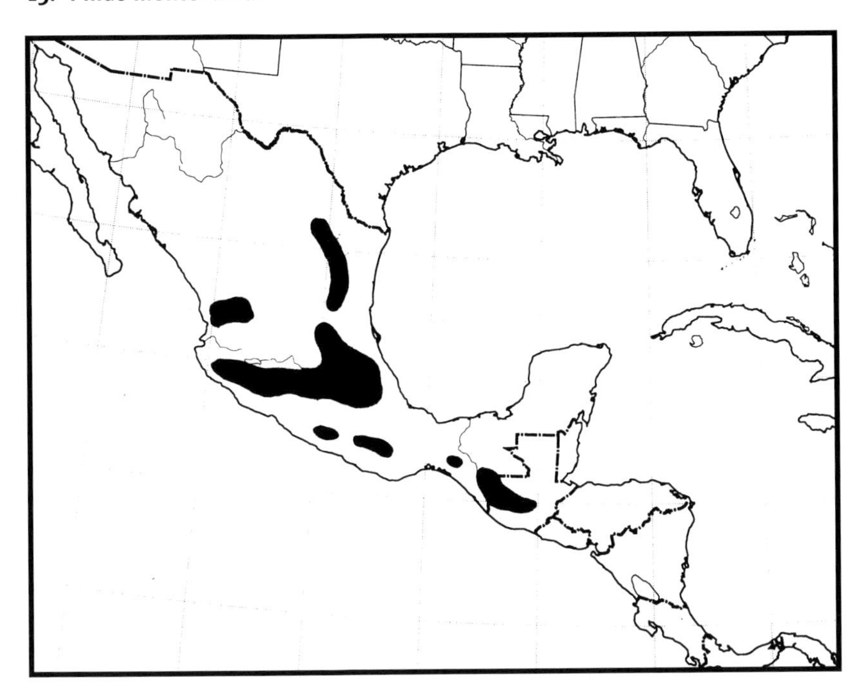

16. Pinus devoniana

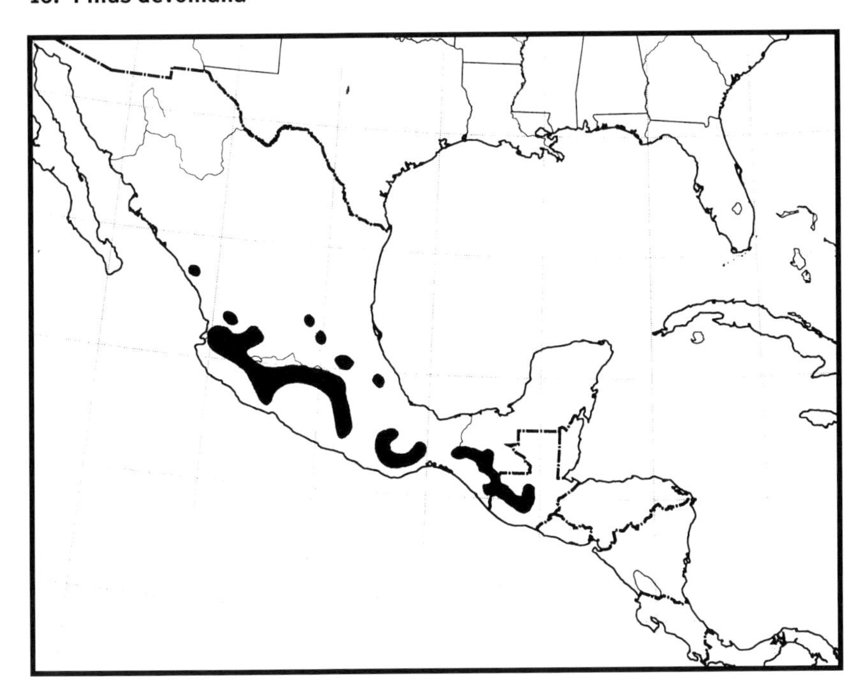

17. **Pinus douglasiana**

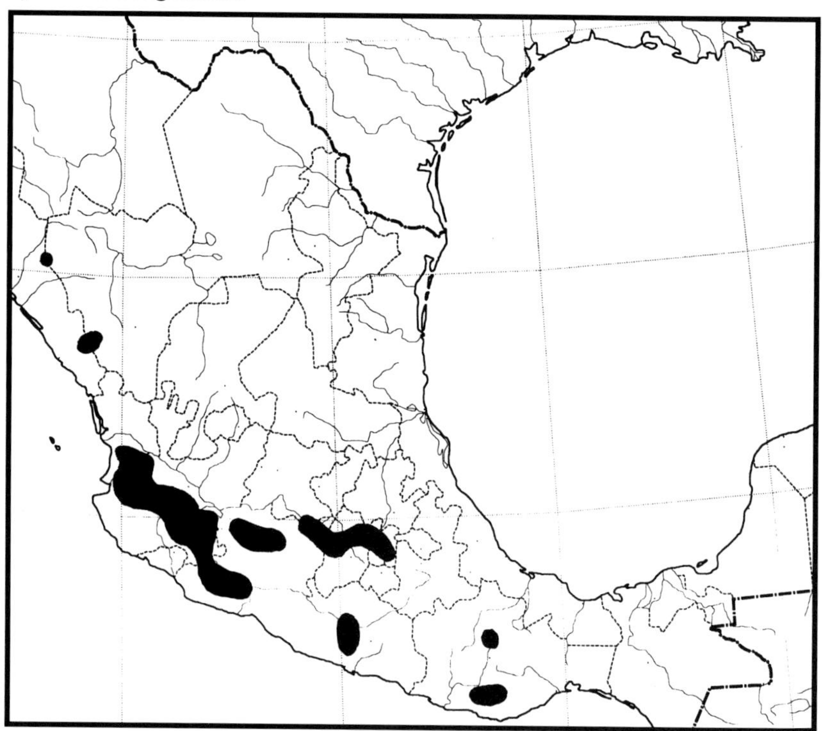

18. **Pinus maximinoi**

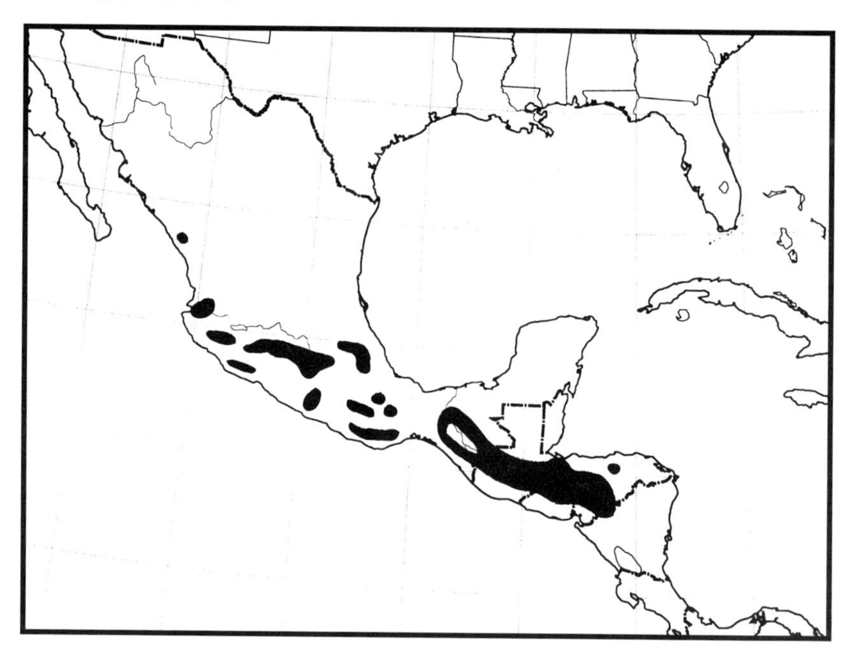

19. Pinus lumholzii

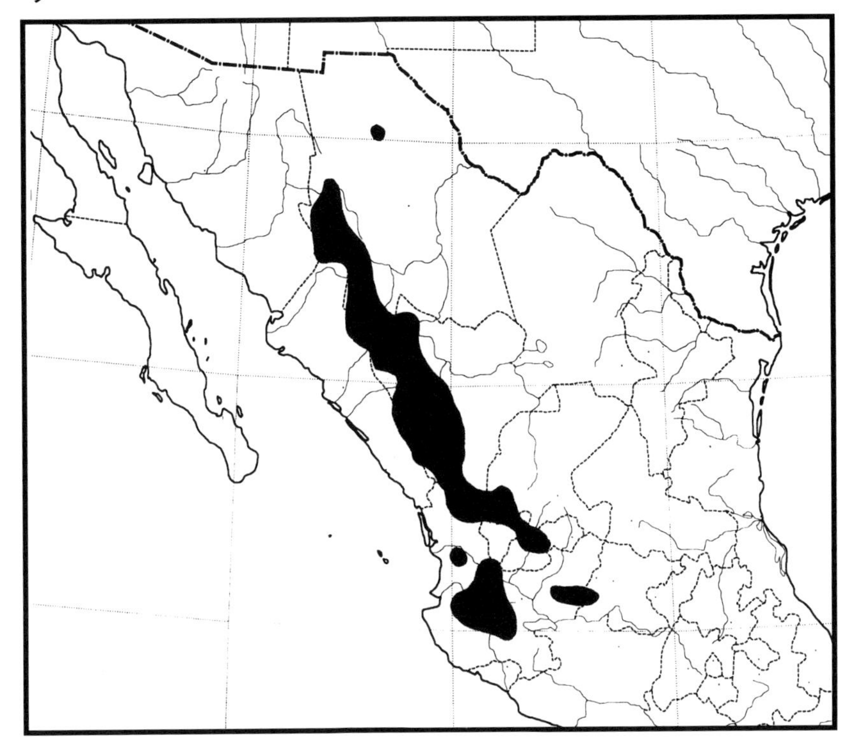

20. Pinus oocarpa

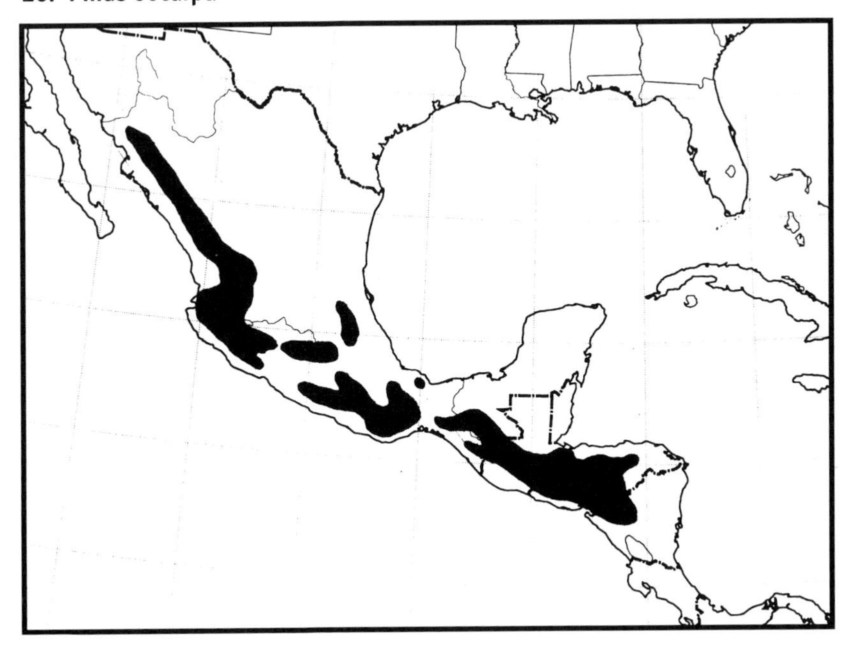

21. Pinus praetermissa

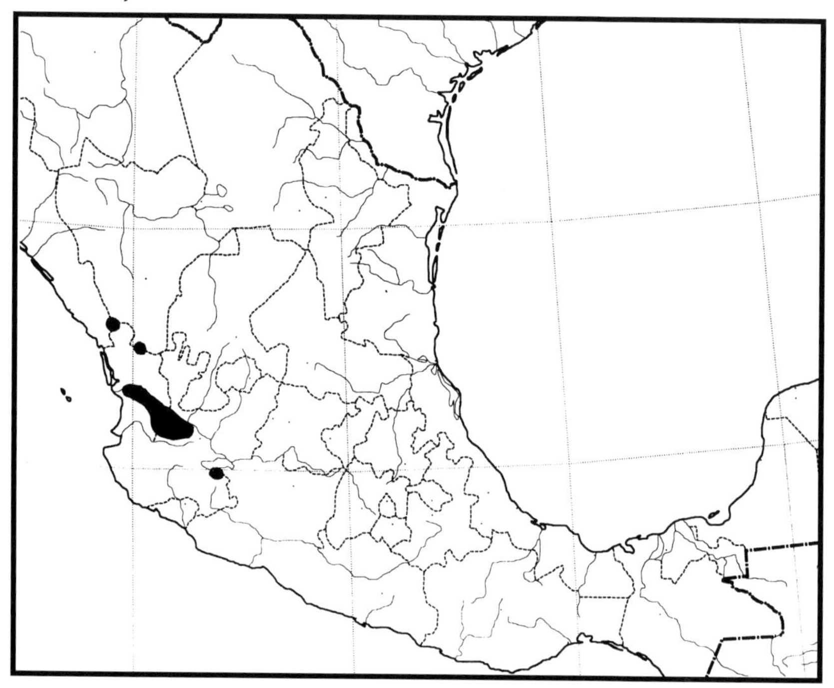

22. Pinus patula

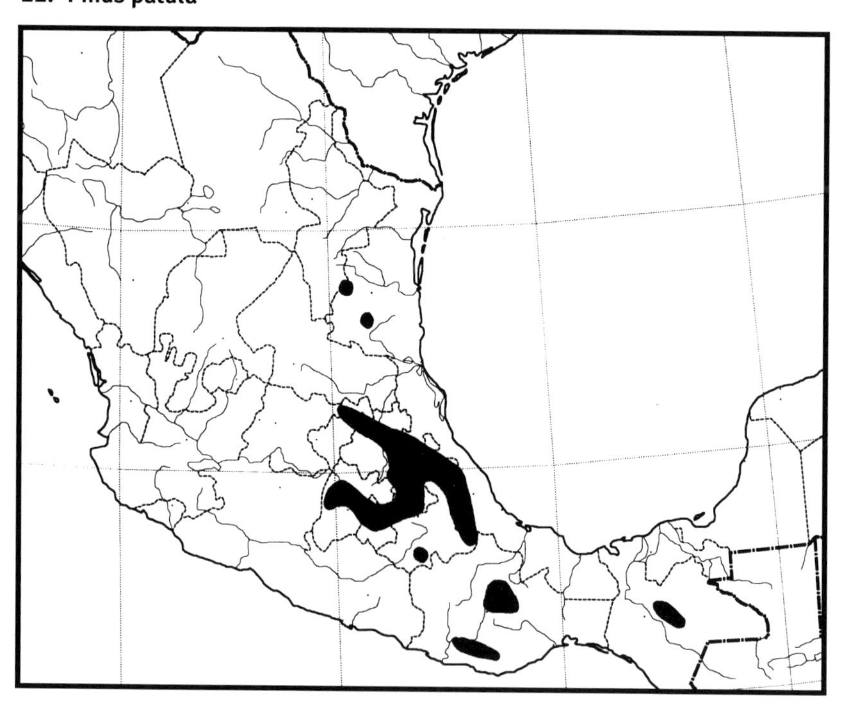

23. Pinus jaliscana

24. Pinus tecunumanii

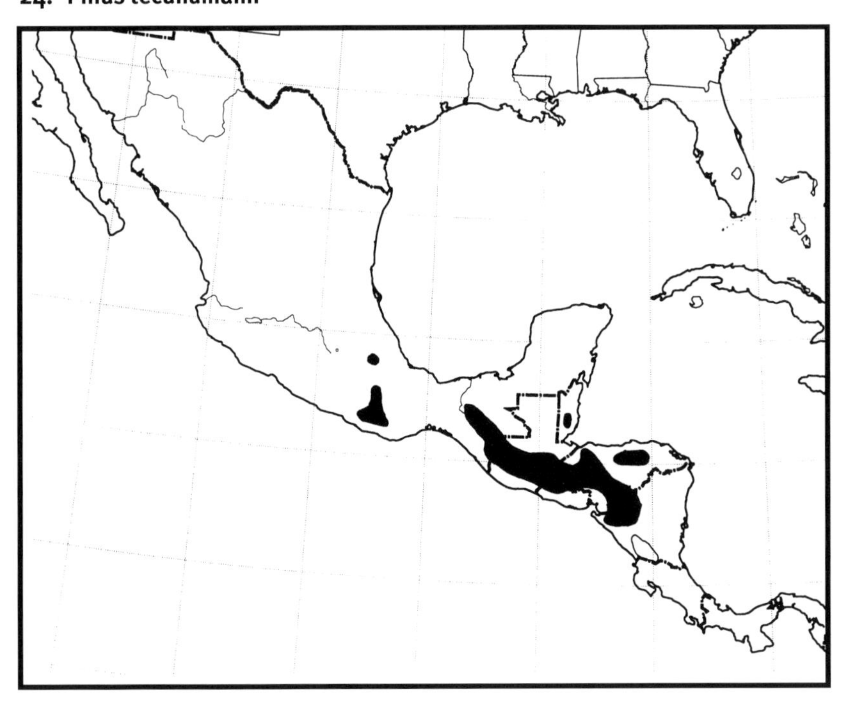

25. Pinus durangensis

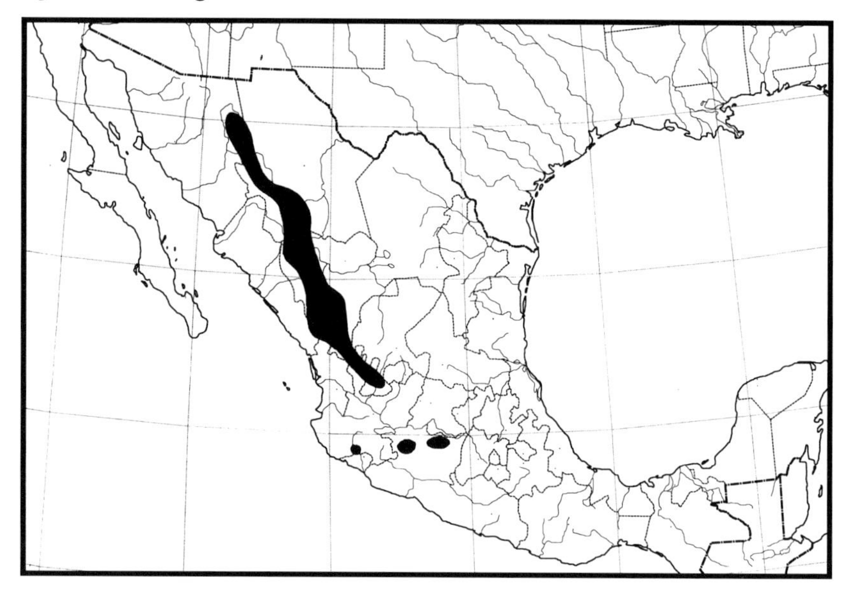

26. Pinus lawsonii

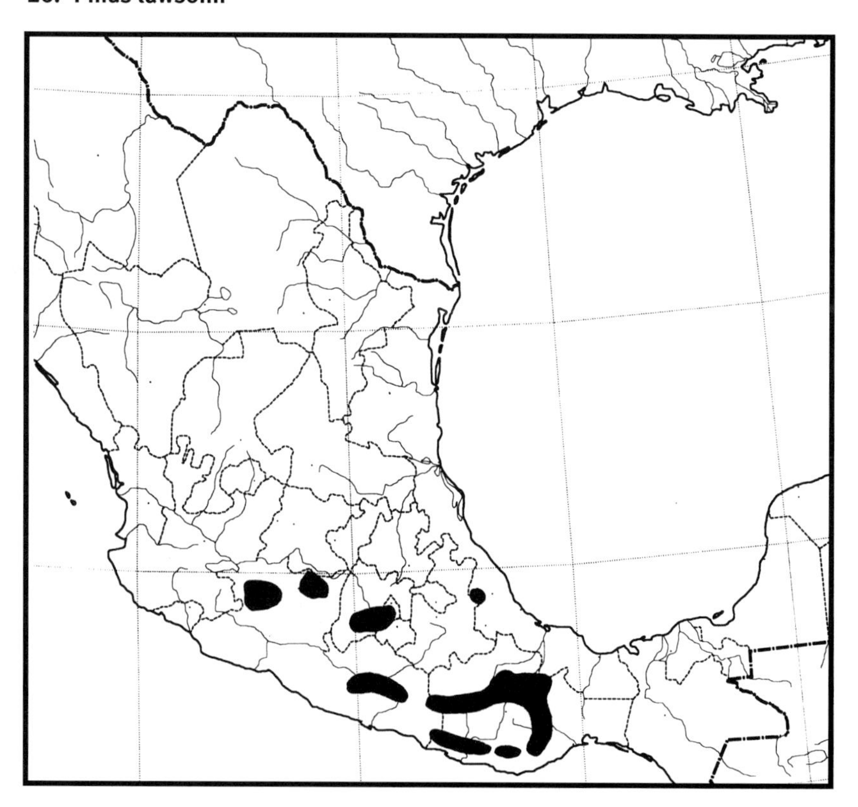

27. Pinus pringlei

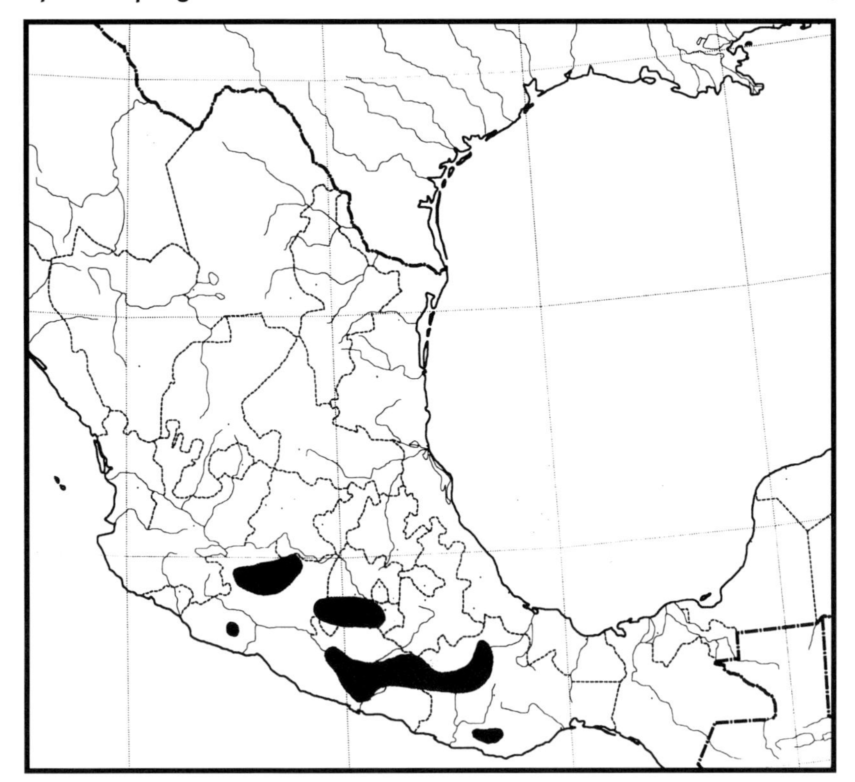

28. Pinus teocote

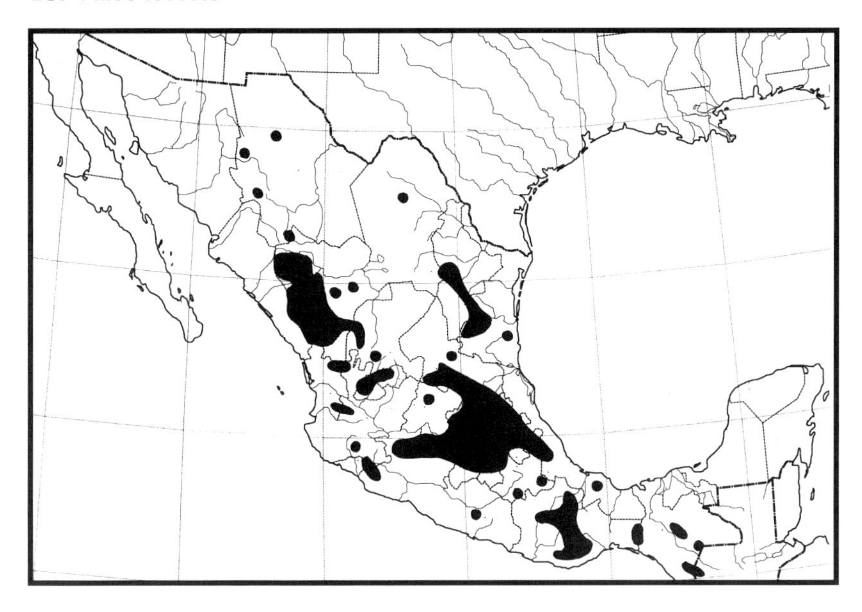

29. Pinus muricata

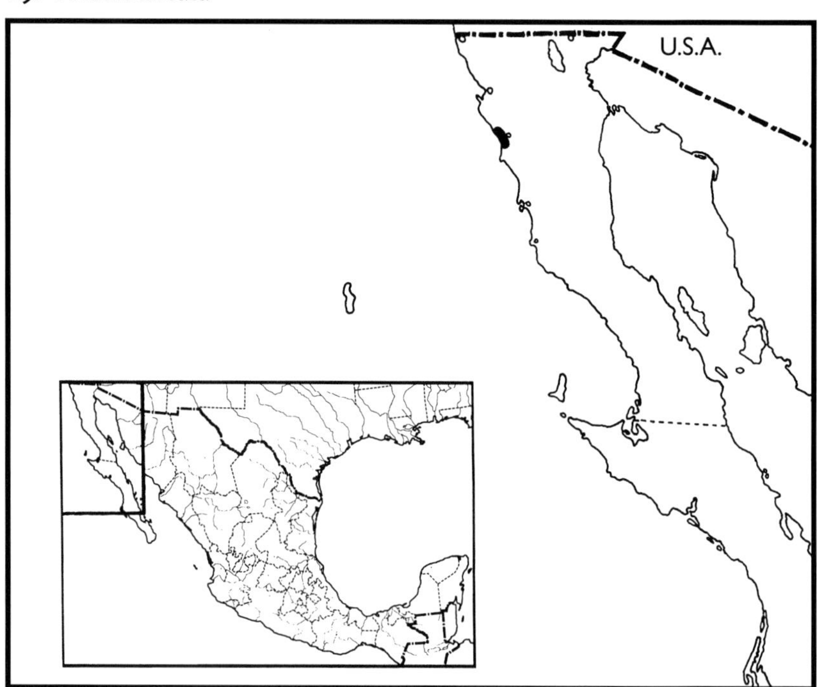

30. Pinus radiata var. **binata**

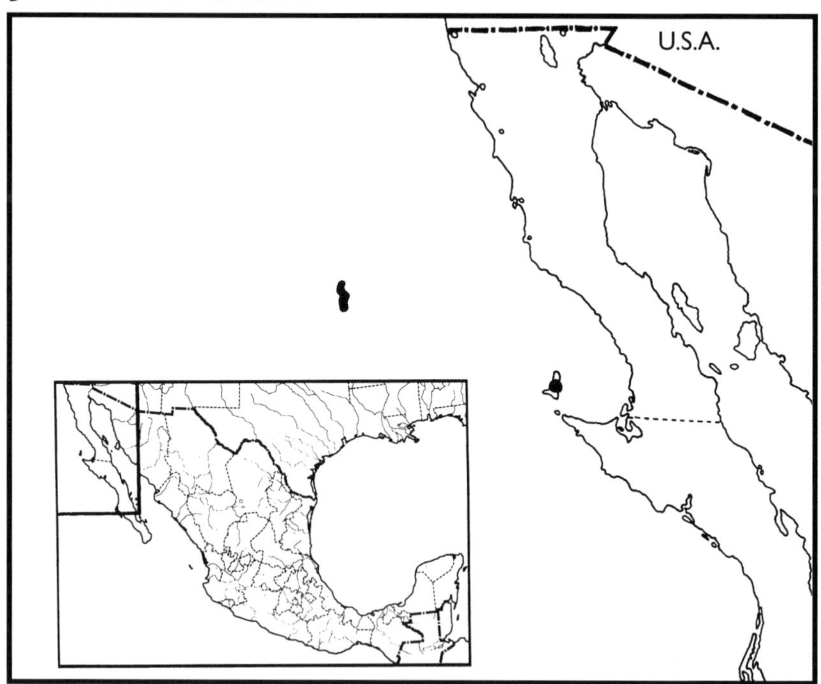

31. Pinus attenuata

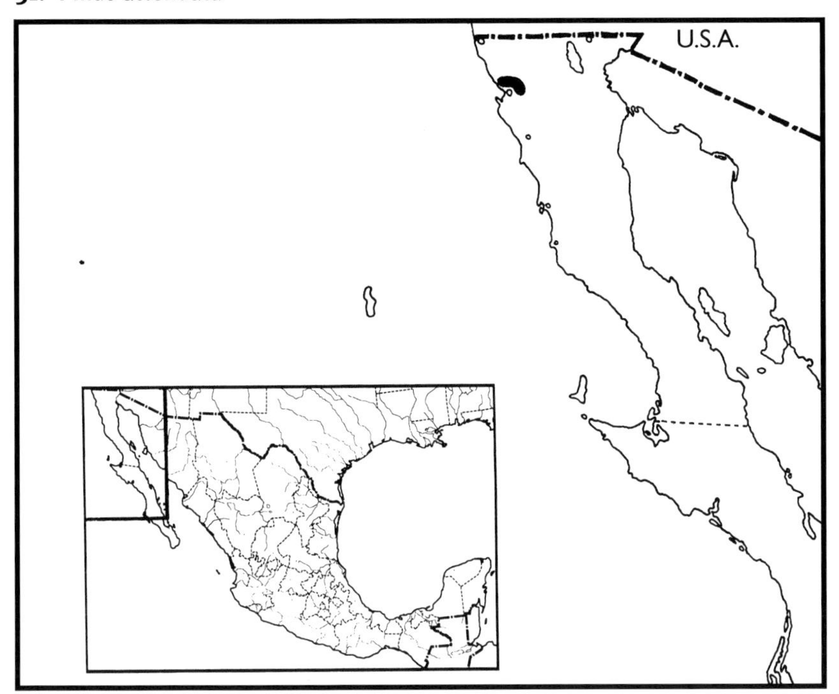

32. Pinus greggii

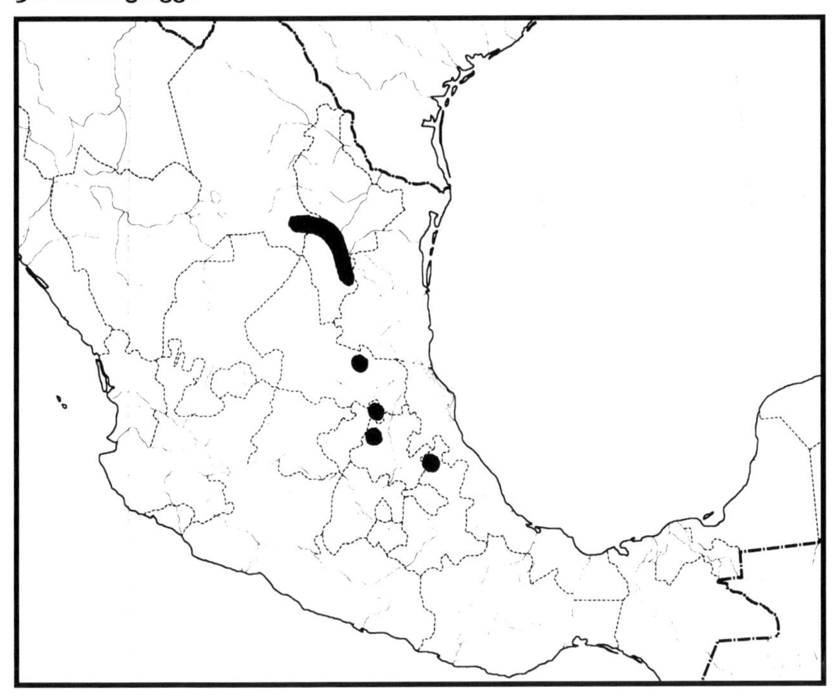

33. Pinus coulteri

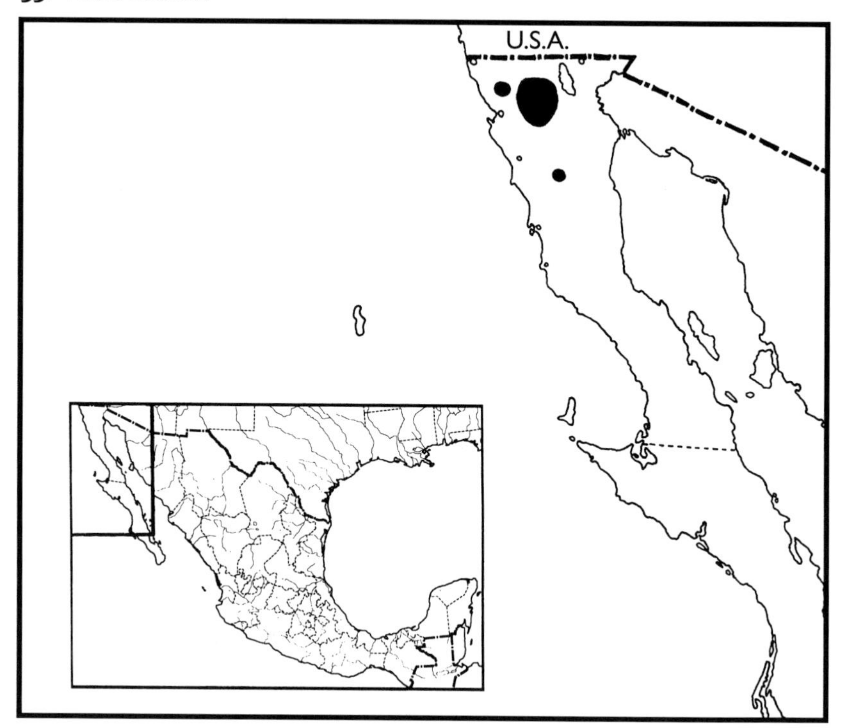

34. Pinus ayacahuite

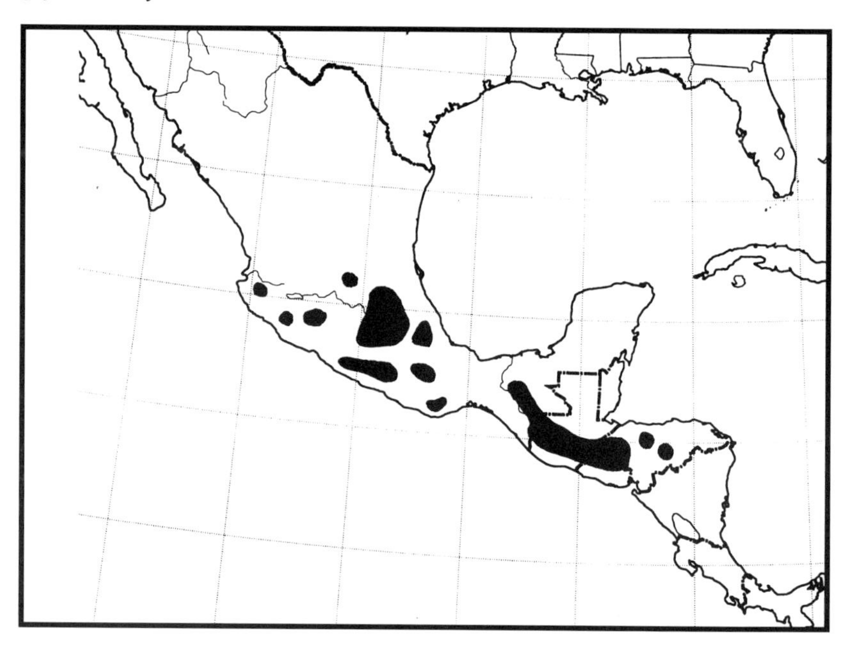

35. Pinus lambertiana

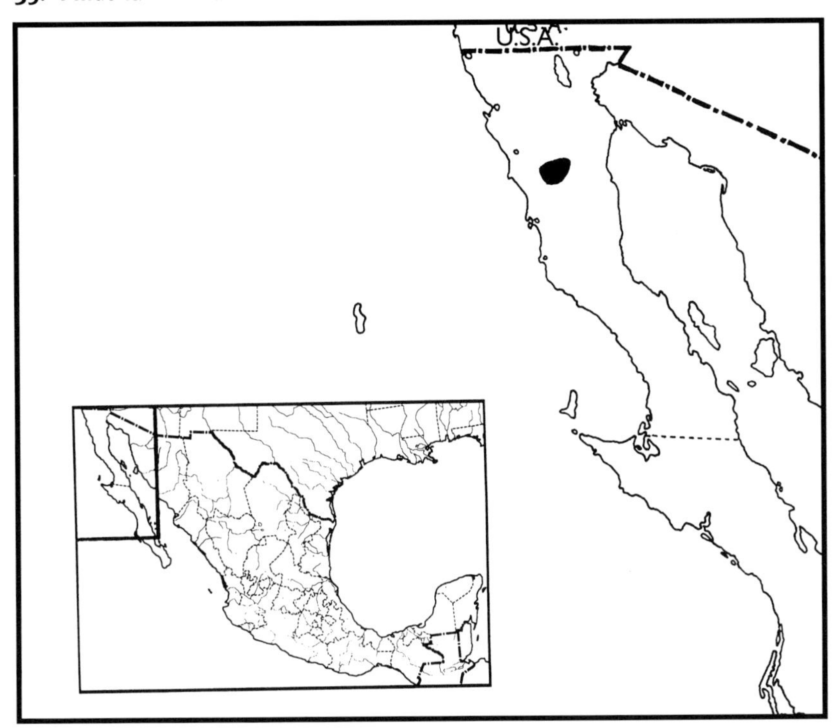

36. Pinus flexilis var. **reflexa**

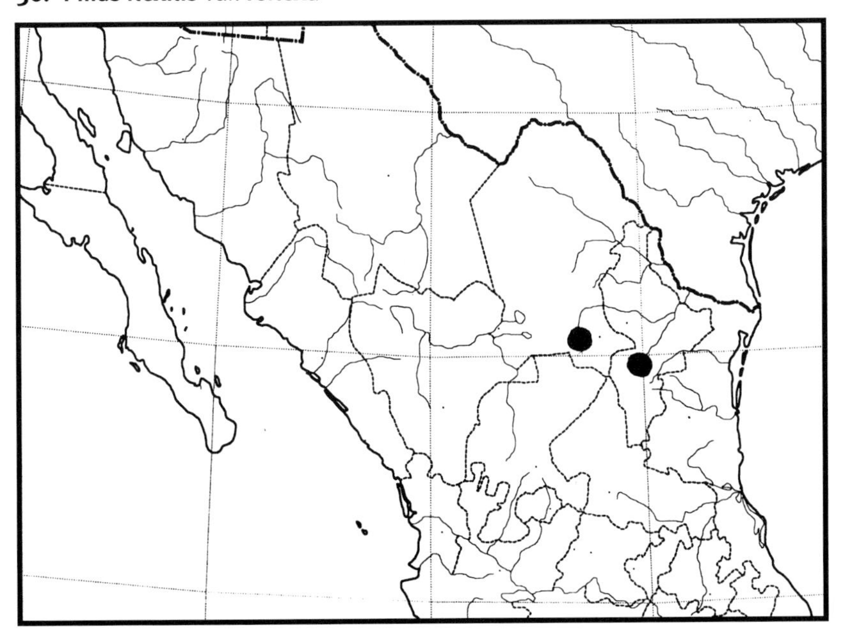

37. Pinus strobiformis

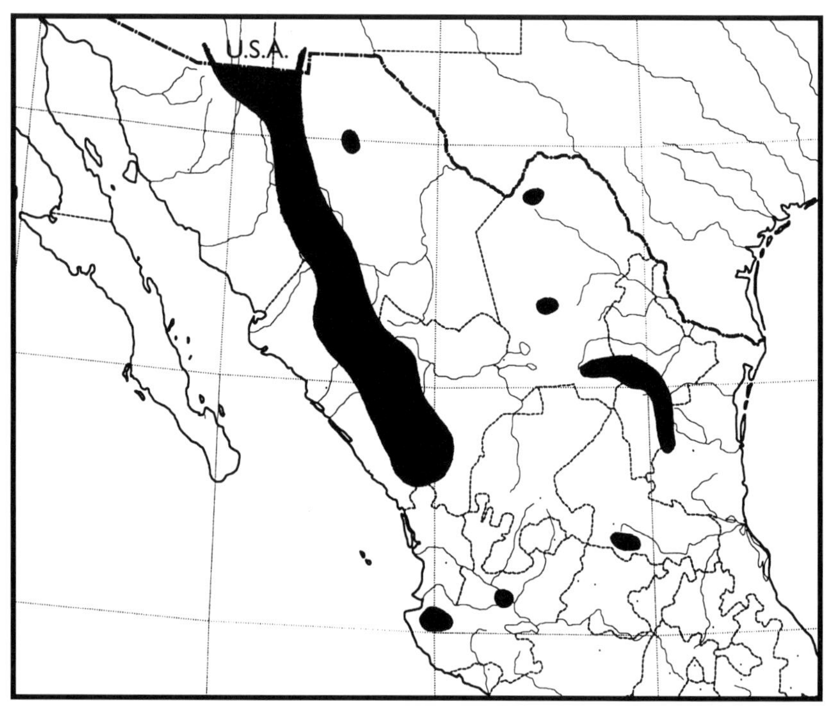

38. Pinus strobus var. **chiapensis**

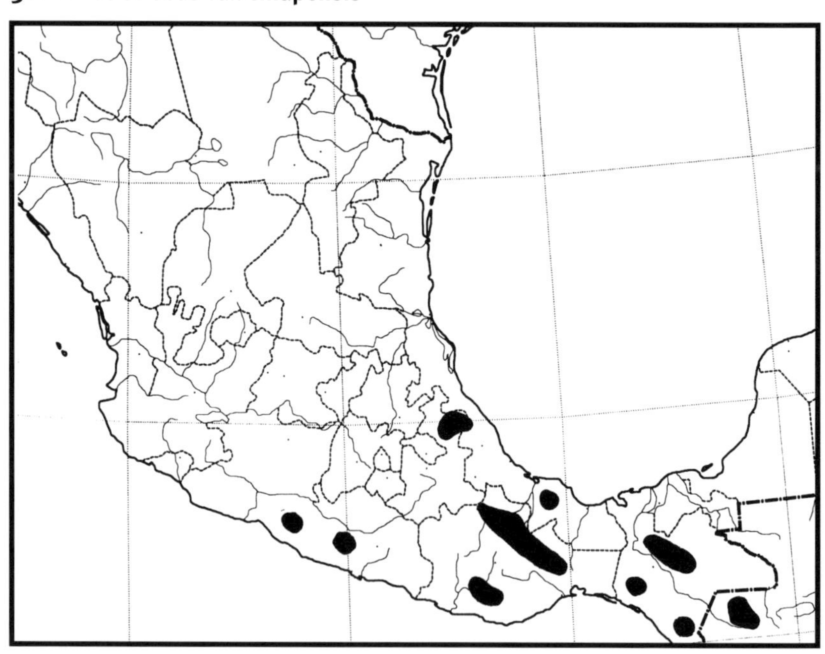

39. Pinus rzedowskii

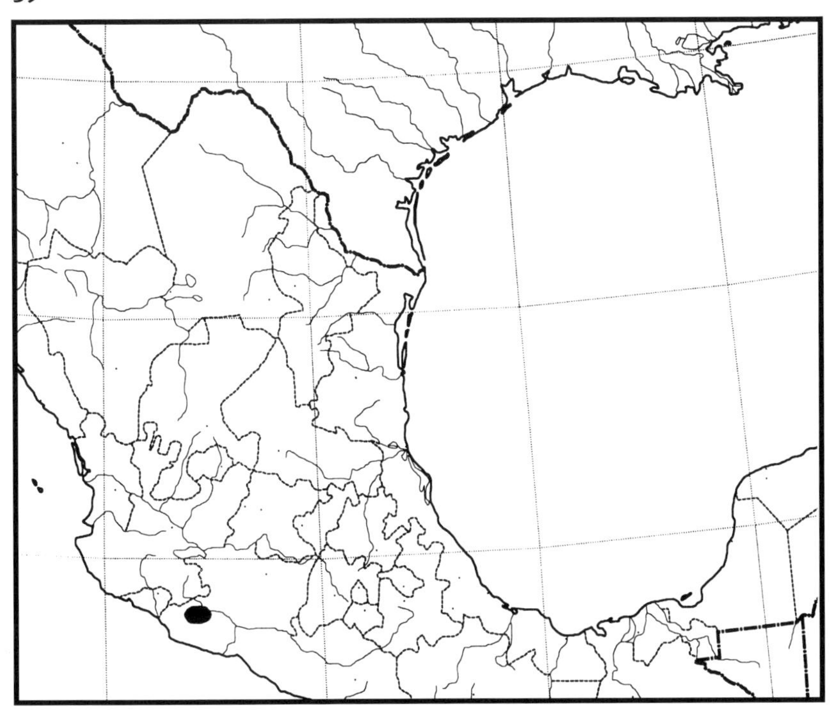

40. Pinus. maximartinezii

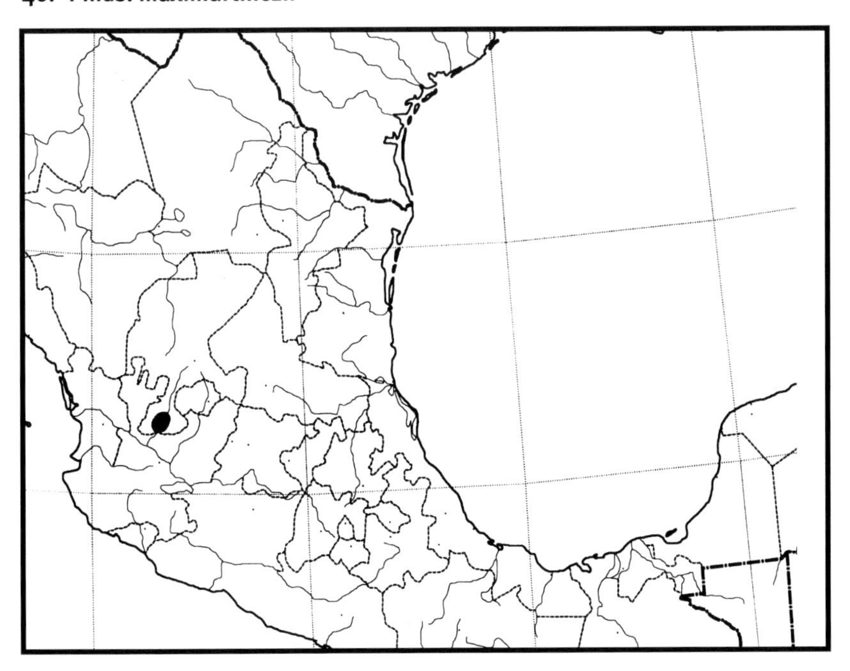

41. Pinus nelsonii

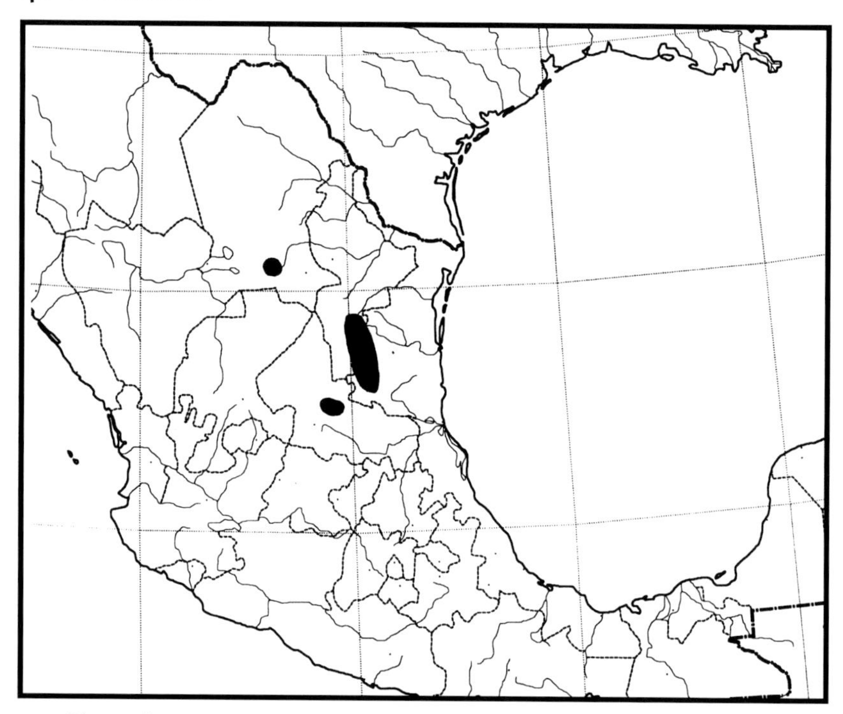

42. Pinus pinceana

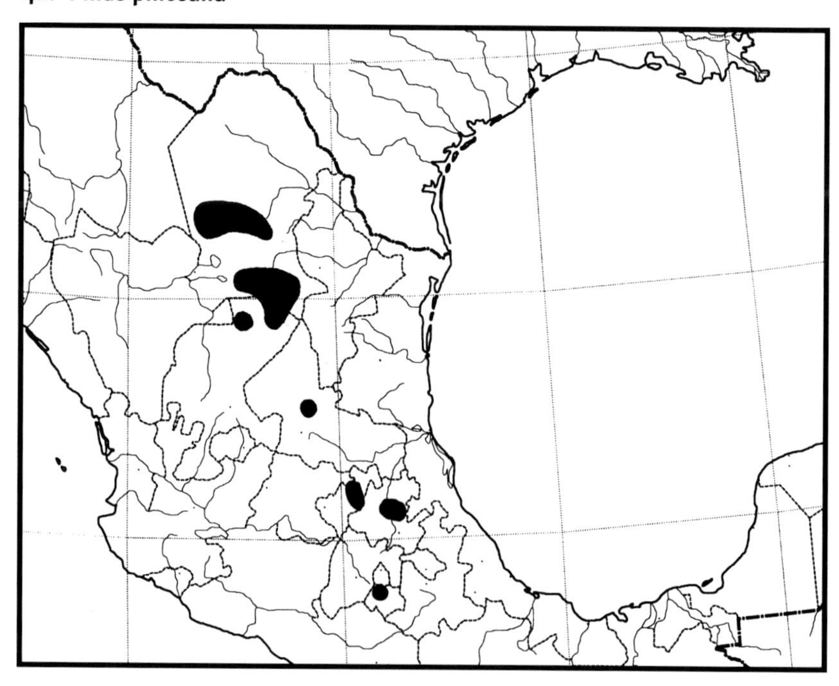

43. **Pinus cembroides**

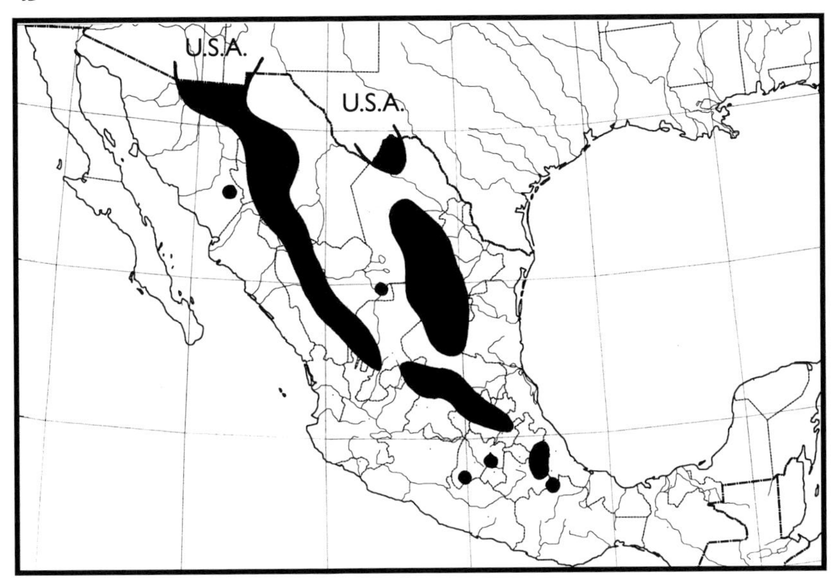

44. **Pinus culminicola**

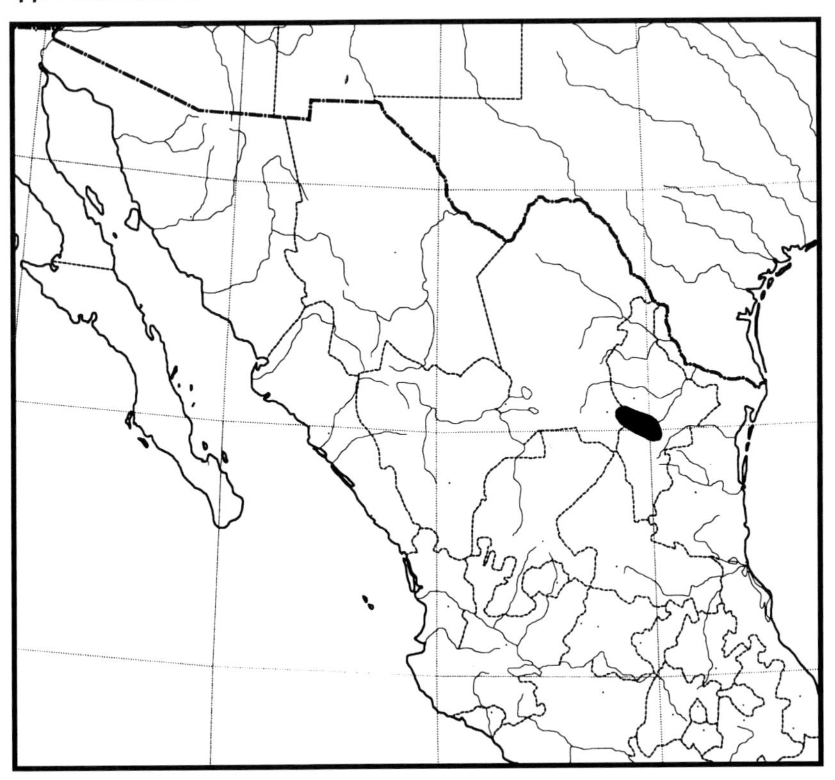

45. Pinus remota

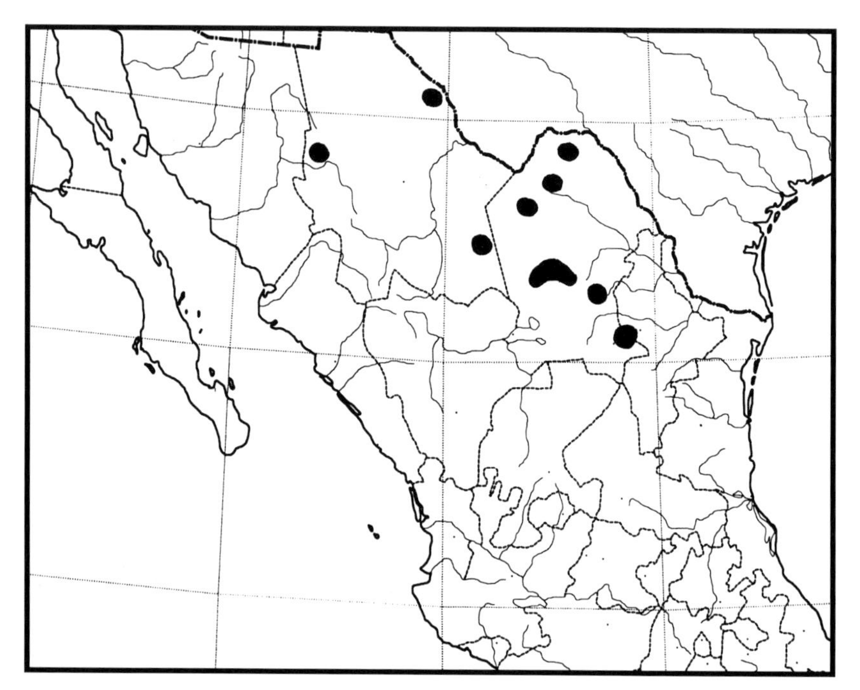

46. Pinus monophylla

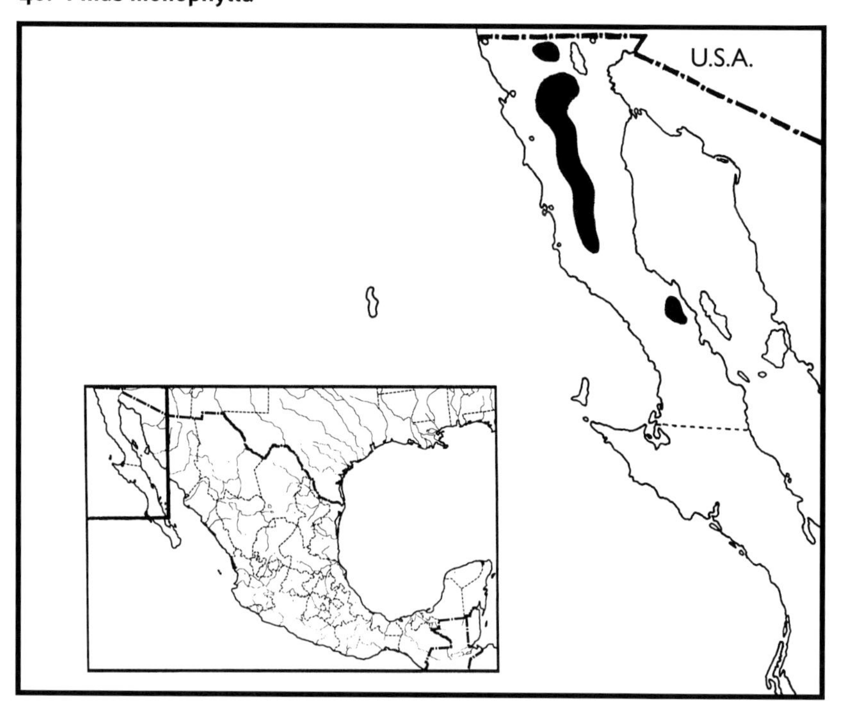

47. Pinus quadrifolia

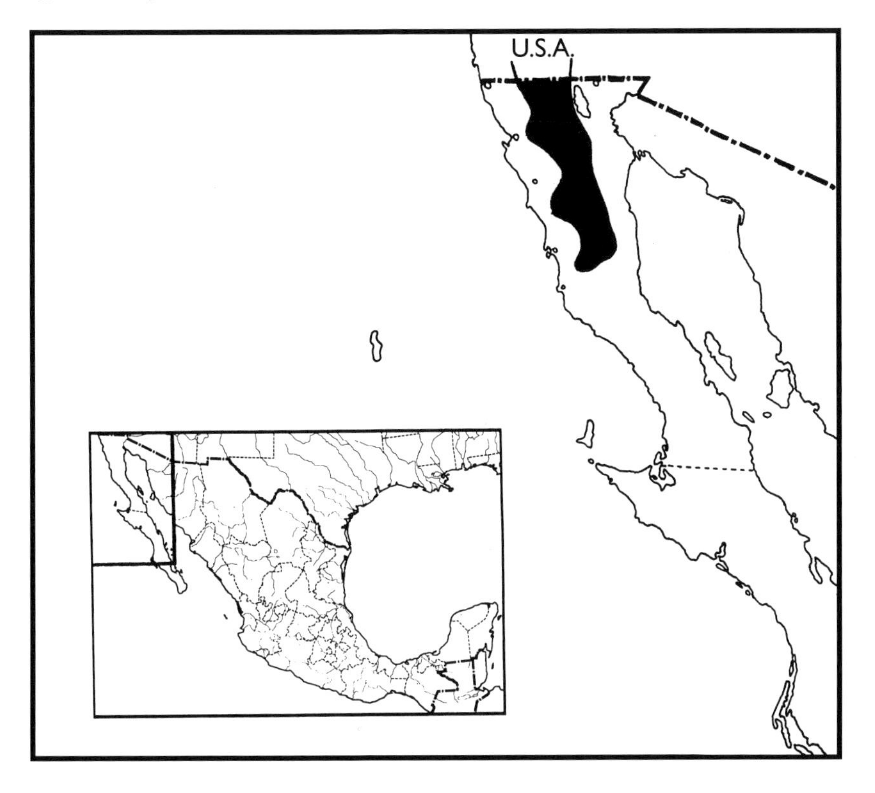